Joint IPPP/CI/ICFA Workshop on Advanced QED Methods for Future Accelerators 2009

Journal of Physics: Conference Series Volume 198

Warrington, United Kingdom
3-4 March 2009

ISBN: 978-1-61738-279-6

Printed from e-media with permission by:

Curran Associates, Inc.
57 Morehouse Lane
Red Hook, NY 12571

Some format issues inherent in the e-media version may also appear in this print version.

Printed by Curran Associates, Inc. (2010)

For permission requests, please contact the Institute of Physics
at the address below.

Institute of Physics
Dirac House, Temple Back
Bristol BS1 6BE UK

Phone: 44 1 17 929 7481
Fax: 44 1 17 920 0979

techtracking@iop.org

Additional copies of this publication are available from:

Curran Associates, Inc.
57 Morehouse Lane
Red Hook, NY 12571 USA
Phone: 845-758-0400
Fax: 845-758-2634
Email: curran@proceedings.com
Web: www.proceedings.com

TABLE OF CONTENTS

The joint IPPP Durham/Cockcroft Institute/ICFA workshop on advanced QED methods for future accelerators took place at the Cockcroft Institute in early March 2009. The motivation for the workshop was the need for a detailed consideration of the physics processes associated with beam–beam effects at the interaction points of future high-energy electron-positron colliders.

There is a broad consensus within the particle physics community that the next international facility for experimental high-energy physics research beyond the Large Hadron Collider at CERN should be a high-luminosity electron–positron collider working at the TeV energy scale. One important feature of such a collider will be its ability to deliver polarised beams to the interaction point and to provide accurate measurements of the polarisation state during physics collisions. The physics collisions take place in very dense charge bunches in the presence of extremely strong electromagnetic fields of field strength of order of the Schwinger critical field strength of 4.4×10^{13} Gauss. These intense fields lead to depolarisation processes which need to be thoroughly understood in order to reduce uncertainty in the polarisation state at collision.

To that end, this workshop reviewed the formalisms for describing radiative processes and the methods of calculation in the future strong-field environments. These calculations are based on the Furry picture of organising the interaction term of the Lagrangian. The means of deriving the transition probability of the most important of the beam–beam processes – Beamsstrahlung – was reviewed. The workshop was honoured by the presentations of one of the founders, V N Baier, of the 'Operator method' – one means for performing these calculations.

Other theoretical methods of performing calculations in the Furry picture, namely those due to A I Nikishov, V I Ritus *et al*, were reviewed and intense field quantum processes in fields of different form – namely those present in intense lasers – were also presented. Within the Furry picture the lowest order physics processes are represented by one vertex Feynman diagrams. Additionally, higher order processes in the Furry picture are thought to be important and are still not fully studied.

The Advanced QED methods workshop also benefited greatly from reports on ongoing and planned experimental work on quantum processes in intense external fields. Some of the experiments reviewed were the NA43 and NA63 experiments using the inter atomic fields in aligned crystals at CERN. In the past, evidence has been obtained from successful experiments using an intense laser at the SLAC experiment E144. The possibility now exists for new experiments with intense laser light with the planned XFEL at DESY and the European Extreme Light Infrastructure.

For upcoming accelerator projects, computer simulations of the first order processes in the Furry Picture during the bunch-bunch collision are being performed using the programs CAIN and Guinea-Pig++. The implementation of spin dynamics in these simulation programs was reported on at the workshop.

This relatively small workshop generated a very productive intermix of theoretical, experimental and computational developments covering this important field of physics. Fruitful discussions took place covering improvements to the models, estimations of the remaining theoretical uncertainties and future updates to the existing simulations. It was felt that ongoing workshops in the same field would be of benefit to all those involved.

The organisers would like to express their sincere thanks to all of the attendees for their contributions, to the staff of the Cockcroft Institute for hosting the workshop, to the IPPP at Durham for providing substantial funding and administrative support, and to ICFA for their sponsorship. We would also like to thank IOP Publishing for their assistance in publishing our proceedings in the *Journal of Physics: Conference Series*.

All papers published in this volume of *Journal of Physics: Conference Series* have been peer reviewed through processes administered by the proceedings Editors. Reviews were conducted by expert referees to the professional and scientific standards expected of a proceedings journal published by IOP Publishing.

Journal of Physics: Conference Series **198** (2009) 012001 doi:10.1088/1742-6596/198/1/012001

Workshop summary: Advanced QED methods for future accelerators

Stefan Hesselbach

IPPP, Durham University, Durham, DH1 3LE, UK

E-mail: `stefan.hesselbach@durham.ac.uk`

Abstract. The talks presented at the workshop "Advanced QED methods for future accelerators" are briefly summarised. They focused on the theoretical description of Intense Field QED (IFQED), the calculation and simulation of IFQED processes and other IFQED applications.

1. Introduction

The workshop "Advanced QED methods for future accelerators" had essentially two aims. One was to gain a firmer understanding of available QED techniques necessary for describing processes in the extreme environments expected during bunch-bunch collisions at future high-luminosity TeV-scale electron-positron accelerators. The other was to gain an analogous understanding in connection with the extreme environments created by high-intensity laser beams. Urgent further attention must be given to the impact of quantum effects on the design of such machines and systems. This workshop was dedicated to reviewing the formalisms for describing these radiative processes and the methods of calculation in the strong-field environments. It was jointly sponsored by the Cockcroft Institute of Accelerator Science and Technology, UK, and the Institute for Particle Physics Phenomenology, University of Durham, UK, and it was a workshop of the series of mini-workshops of the Beam Dynamics Group of ICFA, the International Committee for Future Accelerators. The organisers hope that this workshop was the first in a series and plan a second workshop of this kind in 2010.

In the following, the talks presented at the workshop will be briefly summarised. The methods introduced to describe Intense Field QED (IFQED) phenomena include the quasiclassical operator method of Baier and Katkov and the semiclassical method of Nikishov and Ritus, based on the Furry representation. IFQED calculations for intense laser-light fields were presented as well as the implementation of IFQED processes in computer simulations for high energy accelerators. Further applications include experiments with strong fields and recycled accelerator parts, lepton pair creation in strong laser fields, strong field QED effects in crystals and geometric and ultrarelativistic methods.

2. Intense Field QED methods

Intense-field phenomena can be befittingly described on the quantum mechanical level within the Furry or bound-interaction representation [1], and in [2] an introduction to this approach was given. The Furry representation is a variant of the interaction representation intermediate between the Heisenberg and Dirac representations. In this representation the effects of the external field are included in the wave functions and the eigenstates describe a bound system instead of a free particle system. Hence the commutation relations, the gauge transformations and the charge conjugation operation on the electron propagator have to be adapted accordingly and the contributions of self-closed diagrams in the vacuum polarization have to be included.

In [3] an introduction to the quasiclassical operator method of Baier and Katkov [4] is given and it describes recent developments for the application of this method to pair creation processes by a photon in an external field and for radiation processes in oriented crystals. Furthermore, applications for the description of the superposition of a plane wave and a constant field, radiation processes occurring during beam-beam interaction at linear colliders and the description of inhomogeneous fields are reviewed.

In contribution [5] the derivation of the transition rate of the beamstrahlung process is reviewed, comparing the quasiclassical operator method with the semi-classical method of Nikishov and Ritus [6, 7] which uses the Furry representation. The former assumes a classical interaction between the electron and the external field, but a quantum interaction to produce the radiated photon and it is only expected to be valid if the electron is ultra-relativistic before and after the photon emission. The latter is a full calculation within the usual S-matrix theory. Here, there are no kinematic approximations and radiative corrections within the Furry representation have to be included.

3. IFQED calculations

Intense laser light is an important tool for probing the quantum vacuum and other QED effects at high fields. Related IFQED calculations were presented in several talks at the workshop. For example, the QED vacuum can be analysed using light-by-light diffraction [8] whereas other related phenomena of QED in strong light fields include photon merging in light-proton collisions [9], photon splitting in a laser-light field [10], Delbrück scattering in combined Coulomb and laser-light fields [11], laser-light-assisted bremsstrahlung [12] and laser-light-induced channelling of electron positron pairs [13, 14].

In this context, contribution [15] describes Compton scattering at high intensities occurring in collisions of an electron beam with a high-power laser beam. After an overview of this process the expected intensity signatures of the emitted photons are reviewed [16]. Current and planned laser facilities which could be utilised are listed, e.g. the Vulcan facility [17] and the Extreme Light Infrastructure (ELI) project [18], as well as planned experiments at Daresbury laboratory and Forschungszentrum Dresden-Rossendorf.

4. Simulation of IFQED processes in colliders

Contribution [19] reviews the implementation of depolarization effects in the computer code GUINEA-PIG++ [20] which simulates aspects of the beam-beam interaction. In particular, spin-dynamical effects like precession and spin-flip have been included and compared with the implementation in CAIN [21]. Some differences occur which are assumed to arise from a different choice of the coordinate systems: in CAIN the z-direction is the direction of the incoming particle whereas GUINEA-PIG++ uses a fixed z-direction. Despite these differences good overall agreement is found for the main beam-parameter sets of the ILC.

A further talk summarised the implementation and limitations of IFQED processes in CAIN and contained suggestions for improving these, focusing especially on second order pair-production processes. The Volkov solutions for these pair-production processes were discussed as well as the two vertex auxiliary functions and the Volkov propagator poles. Results of recent simulations using CAIN can be found in [22].

5. IFQED applications

Contribution [23] explains how low energy experiments employing intense laser beams or strong electric and magnetic fields can be used to search for phenomena beyond the Standard Model, focusing on models with minicharged particles and hidden photons. This can be achieved with dark current experiments in accelerator cavities (AC/DC experiments), by testing Coulomb's law with Cavendish-type experiments, with light-shining-through-a-wall experiments including tunnelling phenomena of the third kind and with laser-beam polarization experiments. Examples for this kind of experiment include the ALPS experiment at DESY and an experiment to search for hidden-sector photons at Daresbury laboratory using an ILC superconducting RF cavity.

Another talk focused on possibilities for lepton pair creation in strong laser-light fields. For example, in laser-driven colliders [24] muon pair production from positronium atoms can be probed [25]. Pair production can also occur, via absorption of multiple photons, when a beam of relativistic particles collides with a pulse of laser light [26]. Free pair creation in laser light-nucleus collisions [27] is also possible. For these, petawatt lasers or UV/X-ray lasers (e.g. DESY FLASH) and ion accelerators at CERN or GSI could be utilised. Pair creation can also take place in counter-propagating laser beams which create standing electromagnetic waves where pronounced magnetic-field effects occur [28].

In contribution [29] the connection between IFQED effects in crystals [30] and in beamstrahlung is presented. Experiments carried out with crystals can be used to support beamstrahlung calculations for future linear colliders. The results of experimental studies of quantum suppression of synchrotron radiation, coherent pairs and the trident process confirm the validity of the theory in the strong field regime.

In a further talk geometric and ultrarelativistic methods for accelerator physics were introduced. In the ultrarelativistic approach the nonlinear system of partial differential equations which describes the particles in a beam is replaced by a hierarchy of linear equations. Examples for the ultrarelativistic expansion for a cold fluid, for a single charged particle [31] and of the Maxwell-Vlasov equations [32] were presented.

6. Summary

The workshop "Advanced QED methods for future accelerators" provided reviews of the quasiclassical operator method of Baier and Katkov and the semi-classical method of Nikishov and Ritus, which is based on the Furry representation.

In anticipation of future high-luminosity electron-positron accelerators at the TeV-scale, the implementation of IFQED processes in the simulation of the beam-beam interaction in the codes GUINEA-PIG++ and CAIN was summarised.

For the application of IFQED in intense laser-light fields several different calculations and related ongoing and planned experiments were described, including vacuum polarization effects in intense laser-light fields, IFQED effects in Compton scattering and lepton pair creation in strong laser-light fields. Further talks reviewed strong fields in recycled accelerator parts as a laboratory for fundamental physics, experiments on strong field QED in crystals and geometric and ultrarelativistic methods for accelerator physics.

References

[1] Furry W H 1951 *Phys. Rev.* **81** 115–24
[2] Moortgat-Pick G 2009 *These proceedings*
[3] Baier V N and Katkov V M 2009 *These proceedings*
[4] Baier V N and Katkov V M 1967 *Sov. Phys. JETP* **25** 944–7
[5] Hartin A 2009 *These proceedings*
[6] Nikishov A I and Ritus V I 1964 *Sov. Phys. JETP* **19** 529–41
[7] Nikishov A I and Ritus V I 1964 *Sov. Phys. JETP* **19** 1191–9
[8] Di Piazza A, Hatsagortsyan K Z and Keitel C H 2006 *Phys. Rev. Lett.* **97** 083603 (*Preprint* hep-ph/0602039)

[9] Di Piazza A, Hatsagortsyan K Z and Keitel C H 2008 *Phys. Rev.* A **78** 062109 (*Preprint* arXiv:0906.5576)
[10] Di Piazza A, Milstein A I and Keitel C H 2007 *Phys. Rev.* A **76** 032103
[11] Di Piazza A and Milstein A I 2008 *Phys. Rev.* A **77** 042102
[12] Lötstedt E, Jentschura U D and Keitel C H 2007 *Phys. Rev. Lett.* **98** 043002
[13] Lötstedt E, Jentschura U D and Keitel C H 2008 *Phys. Rev. Lett.* **101** 203001
[14] Lötstedt E, Jentschura U D and Keitel C H 2009 *New J. Phys.* **11** 013054
[15] Heinzl T 2009 *These proceedings*
[16] Harvey C, Heinzl T and Ilderton A 2009 *Phys. Rev.* A **79** 063407 (*Preprint* arXiv:0903.4151)
[17] The Vulcan 10 Petawatt Facility
`http://www.clf.rl.ac.uk/Facilities/vulcan/projects/10pw/10pw_index.htm`
[18] The Extreme Light Infrastructure (ELI) project
`http://www.extreme-light-infrastructure.eu`
[19] Rimbault C, Le Meur G, Blampuy F, Bambade P and Schulte D 2009 *These proceedings*
[20] Le Meur G *et al* 2008 EUROTeV-Report-2008-067
`http://flc.web.lal.in2p3.fr/mdi/BBSIM/bbsim.html`
[21] Yokoya K 2003 *User's Manual of CAIN - Version 2.35*
`http://lcdev.kek.jp/~yokoya/CAIN/cain235/CainMan235.pdf`
[22] Bailey I R *et al* 2008 *In the Proceedings of 11th European Particle Accelerator Conference (EPAC 08), Magazzini del Cotone, Genoa, Italy, 23-27 Jun 2008,* MOPP024, EUROTeV-Report-2008-026
[23] Jäckel J 2009 *These proceedings* (*Preprint* arXiv:0908.0730)
[24] Hatsagortsyan K Z, Müller C and Keitel C H 2006 *Europhys. Lett.* **76** 29 (*Preprint* physics/0602093)
[25] Müller C, Hatsagortsyan K Z and Keitel C H 2006 *Phys. Rev.* D **74** 074017 (*Preprint* physics/0602106)
[26] Müller C 2009 *Phys. Lett.* B **672** 56 (*Preprint* arXiv:0811.0976)
[27] Müller C, Deneke C and Keitel C H 2008 *Phys. Rev. Lett.* **101** 060402 (*Preprint* arXiv:0803.3512)
[28] Ruf M, Mocken G R, Müller C, Hatsagortsyan K Z and Keitel C H 2009 *Phys. Rev. Lett.* **102** 080402 (*Preprint* arXiv:0810.4047)
[29] Esberg J and Uggerhøj U I 2009 *These proceedings*
[30] Uggerhøj U I 2005 *Rev. Mod. Phys.* **77** 1131–71
[31] Burton D A, Gratus J and Tucker R W 2007 *Annals Phys.* **322** 599–630 (*Preprint* physics/0512218)
[32] Gratus J 2008 *Preprint* arXiv:0809.2944

Joint IPPP, CI and ICFA Workshop on Advanced QED Methods for Future Accelerators IOP Publishing
Journal of Physics: Conference Series **198** (2009) 012002 doi:10.1088/1742-6596/198/1/012002

The Furry Picure

Gudrid Moortgat-Pick

Institute for Particle Physics Phenomenology, Physics Dept., University of Durham, DH1 3LE Durham, UK

Cockcroft Institute of Accelerator Science and Technology, Daresbury Laboratory, Warrington, WA4 4AD Cheshire, U.K.

E-mail: g.a.moortgat-pick@durham.ac.uk

Abstract. A short summary is given on the key points of the Furry picture. It describes quantum mechanical systems in external fields. It differs therefore from the quantum mechanical Schrödinger, Heisenberg and interaction representations.

1. Overview about the different quantum mechanical representations

1.1. Quantum mechanical description

Any quantum mechanical state is given by its state vector $|\Psi>$. Usually these vectors are normalized $<\Psi|\Psi>=1$. The state vector is composed by its different eigenstates:

$$|\Psi>=\sum_i c_i|e_i> \tag{1}$$

Experimental observables are presented by Hermitian operators O

$$O|e_i>=\omega_i|e_i> \tag{2}$$

with the eigenvalues ω_i. The set of all eigenstates build a complete orthonormal basis:

$$<e_i|e_i'>=\delta_{ii'}. \tag{3}$$

The expectation values of the operators correspond to experimental mean values of the observables and are given by:

$$\begin{aligned} <O>=<\Psi|O|\Psi> &= \sum_i\sum_j <\Psi|e_i><e_i|A|e_j><e_j|\Psi> &(4)\\ &= \sum_i\sum_j \omega_i <\Psi|e_i><e_i|e_j><e_j|\Psi> &(5)\\ &= \sum_i \omega_i|<e_i|\Psi>|^2 &(6)\end{aligned}$$

Two observables are simultaneously measurable if they commute:

$$[A,B]|\Psi>=0|\Psi>=A(B|\Psi>)-B(A|\Psi>). \tag{7}$$

Journal of Physics: Conference Series **198** (2009) 012002 doi:10.1088/1742-6596/198/1/012002

The correspondence principle guarantees that the classical physics is obtained in the limit of large statistics. The quantum system is described via the same observables as in classical mechanics. However, these observables are functions of the space and momentum operators X_i, P_j which correspond to the generalized conjugated coordinates x_i, p_j.

The momentum and space operators fulfill the Heisenberg commutation relations:

$$[x_i, x_j] = 0 \quad (8)$$
$$[p_i, p_j] = 0 \quad (9)$$
$$[x_i, p_j] = i\hbar\delta_{ij} \quad (10)$$
$$(11)$$

The commutation relations eqn.(8)–(10) lead directly to the quantum mechanical uncertainty relation:

$$\Delta X \cdot \Delta P \leq \frac{\hbar}{2} \quad (12)$$

Space and momentum of a system are not simultaneously measurable.

There exist two presentations, the space– and the momentum representation. Measuring, for instance, the momentum spectrum of a quantum system and taking into account the uncertainty relation it is plausible that a continuous spectrum rather than a discrete spectrum can be expected. The Kronecker δ in the condition for orthonormal eigenstates, eq.3, can therefore be replaced by the δ-function:

$$< \vec{p}|\vec{p'} >= \delta(\vec{p} - \vec{p'}). \quad (13)$$

The δ-distribution is defined as

$$\int_{-\infty}^{+\infty} dx\delta(x - x_0)f(x) = f(x_0) \quad (14)$$

Applying a Fourier transformation, the δ-function can be expressed:

$$\delta(x) = \frac{1}{2\pi\hbar}\int_{-\infty}^{+\infty} dp \exp^{-ipx/\hbar} \quad (15)$$

The summation over all possible eigenstates has to be replace by an integration over the full phase space:

$$\int d^3p|\vec{p} >< \vec{p}| = 1 \quad (16)$$

Analogously the spectrum of space operator can be regarded as being continuous.

The momentum and position operator act on the wave function in the momentum space:

$$X\Psi(\vec{p}) = i\hbar\nabla_{\vec{p}}\Psi(\vec{p}), \quad (17)$$
$$P\Psi(\vec{p}) = \vec{p}\Psi(\vec{p}). \quad (18)$$

The wave function in momentum space is the Fourier transform of the wave function in position space and vice versa:

$$\Psi(\vec{p}) = < \vec{p}|\Psi >= \frac{1}{\sqrt{2\pi\hbar}}\int d^3x \exp^{-i\vec{p}\vec{x}/\hbar}\Psi(\vec{x}) \quad (19)$$
$$\Psi(\vec{x}) = < \vec{x}|\Psi >= \frac{1}{\sqrt{2\pi\hbar}}\int d^3p \exp^{i\vec{p}\vec{x}/\hbar}\Psi(\vec{p}) \quad (20)$$

1.2. Time-dependence in Quantum Mechanics
There exist several possibilities to describe the time development of a quantum mechanical system. Analogous to the classical equations of motion (Newton, Lagrange or Hamilton), one applies a quantum mechanical equation of motions where momentum and position are replaced by their operators. But that requires that also the state vector and the properties of the quantum system on which the operators acts has to be taken into account when discussing the time development. Besides, time is not an observable of the system but just a parameter.

1.2.1. Schrödinger picture In the Schrödinger picture the complete time dependence is presented by the time development of the quantum state vector. The time development is given by an unitary time development operator,

$$|\Psi_S(t)>= U(t,t_0)|\Psi_S(t_0)>, \tag{21}$$

given by the time-dependent Schördinger equation,

$$\frac{d}{dt}|\Psi_S>= \frac{-i}{\hbar}\mathcal{H}|\Psi_S> . \tag{22}$$

The index 'S' denotes the Schrödinger picture. In case it is a closed system, i.e. $\mathcal{H} \neq \mathcal{H}(t)$, the unitary operator is given by

$$U(t,t_0) = \exp^{\frac{i}{\hbar}(t-t_0)\mathcal{H}} . \tag{23}$$

The observables and their eigensystem is constant in time. The mean value is constant if $|\Psi>$ is in an energy eigenstate, i.e. stationary state.

1.2.2. Heisenberg picture In the Heisenberg picture the complete time dependence is governed by the observables, i.e. the operators and their eigensystem:

$$|e_i,t> = U^\dagger(t)|e_i>, \tag{24}$$

$$O_H(t) = U^\dagger(t)O_S U(t), \tag{25}$$

index 'H' denoting the Heisenberg picture and corresponding to the equation of motion:

$$\frac{dO}{dt} = \frac{[O,\mathcal{H}]}{i\hbar} + \frac{\delta O}{\delta t}. \tag{26}$$

The state vector $|\Psi_H>$ is constant. The mean value is constant if observable O is conserved quantity.

1.2.3. Dirac picture ('Interaction picture') The time dependence is split between the state vectors and the observables. The full Hamiltonian is given by

$$\mathcal{H} = \mathcal{H}_0 + \mathcal{V}, \tag{27}$$

where $\mathcal{H}_0$ denotes the free Hamiltonian and $\mathcal{V}$ the contribution due to the interaction. The free Hamiltonian $\mathcal{H}_0$ is time-independent, therefore the time development operator is given by

$$U_0(t,t_0) = \exp^{-\frac{i}{\hbar}(t-t_0)\mathcal{H}_0} . \tag{28}$$

The eigenvalues and eigenstates of $\mathcal{H}_0$ are assumed to be known and given by the free system. Therefore the time dependence of the operators are fixed by $\mathcal{H}_0$,

$$i\hbar\frac{d}{dt}O_D = [O_D,\mathcal{H}_0] + \frac{\delta O}{\delta t}, \tag{29}$$

Journal of Physics: Conference Series **198** (2009) 012002 doi:10.1088/1742-6596/198/1/012002

but time dependence of the states are influenced by the interaction Hamiltonian $\mathcal{V}$. In this picture it is more complicated to derive the time development operator for the states than in the Schrödinger picture. The measurements, i.e. the expectation values, however, have to be independent on the chosen presentation:

$$\begin{aligned} < O > (t) &= < \Psi_S(t)|O_S|\Psi_S(t) > = < \Psi_S(t_0)|U^\dagger O_S U|\Psi_S(t_0) > && (30)\\ &= < \Psi_S(t_0)|U^\dagger U_0 U_0^\dagger O_S U_0 U_0^\dagger U|\Psi_S(t_0) > && (31)\\ &= < \Psi_S(t_0)|U^\dagger U_0 O_D U_0^\dagger U|\Psi_S(t_0) > && (32)\\ &= < \Psi_D(t)|O_D|\Psi_D(t) >, && (33)\end{aligned}$$

with

$$|\Psi_D(t) > = U_0^\dagger(t)|\Psi_S(t) > = U_0^\dagger(t)U(t)|\Psi_S(t=0) > = U_D(t)|\Psi_D(t_0) > \qquad (34)$$

and

$$O_D(t) = U_0^\dagger O_S U_0. \qquad (35)$$

Calculating the explicit time dependence, one derives

$$\begin{aligned} \frac{d}{dt}|\Psi_D(t) > &= \frac{d}{dt}\left[U_0^\dagger(t)U(t)|\Psi_S(t=0) >\right] && (36)\\ &= \frac{i}{\hbar}\left[\mathcal{H}_0 - U_0^\dagger \mathcal{H}\right]U(t)|\Psi_S(t=0) > && (37)\\ &= \frac{i}{\hbar}\left[\mathcal{H}_0 - U_0^\dagger(t)\mathcal{H}U_0(t)\right]|\Psi_D(t) > && (38)\\ &= -\frac{i}{\hbar}\mathcal{V}_D|\Psi_D(t) >, && (39)\end{aligned}$$

i.e. $\mathcal{V}_D$ is given by

$$\mathcal{V}_D = U_0^\dagger(t)\mathcal{V}U_0(t) = \exp^{\frac{i}{\hbar}(t-t_0)\mathcal{H}_0}\,\mathcal{V}\exp^{-\frac{i}{\hbar}(t-t_0)\mathcal{H}_0}. \qquad (40)$$

With eqn.(34) and (39), one derives the equation of motion for the development operator:

$$\frac{d}{dt}U_D = -\frac{i}{\hbar}\mathcal{V}_D(t)U_d(t) \qquad (41)$$

and the chronology of the operators gets crucial. The solution has to fulfill

$$U(t,t_0) = U_0(t,t_0) + \int_{t_0}^{t} dt' \frac{U_0(t,t')VU(t',t_0)}{i\hbar} \qquad (42)$$

and can be solved via stepwise integration. Assuming $U_0 \sim U$ leads to a perturbation series, where the first integration step leads to the Born part.

Note that in the interaction picture the eigenstates of the quantum system are assumed to be states of the free system.

2. Introduction in the Furry representation

Including an external field into the calculation requires an extension of the above mentioned description of a quantum system. The Furry representation [1] is a kind of interaction representation, however the main difference is that the eigenstates are assumed to be states of a bound state, caused by the external field. The Hamiltonian is similarly split as in eq.(27)

with the difference that the external field is added to free Hamiltonian and causes therefore 'new' eigenstates:

$$\mathcal{H} = \mathcal{H}_0 + \mathcal{H}_{\text{ext}} + \mathcal{V} = \mathcal{H}_B + \mathcal{V}, \tag{43}$$

where $\mathcal{H}_{\text{ext}}$ denotes the Hamiltonian of a classical, external field that is independent of time, $\mathcal{H}_B$ the Hamiltonian causing the bound states and $\mathcal{V}$ the contribution of the perturbative interaction term.

The wave functions in the interaction representation and in the Furry picture differ, but are related via a canonical transformation:

$$\Psi_F(x) = M^{-1}\Psi_D(x)M, \quad \text{and} \quad \Psi_F^\dagger(x) = M^-1\Psi_D^\dagger(x)M. \tag{44}$$

However, one has to be aware that Ψ_F and Ψ_D are spanned by a different basis system. Consequently, the commutation relations differ in both representations,

$$\{\Psi_F, \Psi_F^\dagger\} \neq \{\Psi_D, \Psi_D^\dagger\}, \tag{45}$$

since Ψ_F, $\Psi_F^\dagger$ contain the effects of the external field. However, in the asymptotic limit $A^\mu_{ext} \to 0$, the usual commutation relations are recovered:

$$\{\Psi_F^\dagger(x), \Psi_F(x')\} = \sum_i \Psi_{F,i}^\dagger(x)\Psi_{F,i}(x') \stackrel{A^\mu_{ext}\to 0}{\longrightarrow} -i\delta(x - x') \tag{46}$$

2.1. Gauge transformation

Quantum field theory is a invariant under gauge transformation:

$$A^\mu(x) \to A^\mu(x) - \frac{\delta\Lambda(x)}{\delta x^\mu}, \tag{47}$$

$$\Psi(x) \to \exp^{-ie\Lambda(x)}\Psi(x), \quad \text{and} \quad \Psi^\dagger(x) \to \exp^{ie\Lambda(x)}\Psi^\dagger(x), \tag{48}$$

where the scalar function $\Lambda(x)$ has to fulfill $\Box^2\Lambda(x) = 0$.

In a bound-state representation, however, a gauge transformation may affect either the external potential A^μ_{ext}, the interaction potential A^μ or even both. Therefore, *in addition* to eqn.(47), (48), the following transformation have to be performed:

$$A^\mu_{ext}(x) \to A^\mu_{ext} - \frac{\delta\Lambda_{ext}(x)}{\delta x^\mu}, \tag{49}$$

$$\Psi(x) \to \exp^{-ie\Lambda_{ext}}\Psi(x), \quad \text{and} \quad \Psi^\dagger(x) \to \exp^{ie\Lambda_{ext}(x)}\Psi^\dagger(x). \tag{50}$$

2.2. Charge conjugation

Another feature in the Furry picture is the absence of absolute symmetry between the wave functions and their charged conjugated wave functions which leads to physical consequences as, for instance, vacuum polarization, see next section.

The charged conjugated wave functions are defined by the usual relations:

$$\Psi'(x) = C\Psi^\dagger(x), \quad \text{and} \quad \Psi'^\dagger(x) = C^{-1}\Psi(x), \tag{51}$$

where

$$C^{-1}\gamma_\mu C = -\gamma_\mu^T. \tag{52}$$

Since the matrix C commutes with the canonical transformation M, the relation to the operators in the Dirac representation is:

$$\Psi_F'(x) = M^{-1}\Psi_D'(x)M, \quad \text{and} \quad \Psi_F'^\dagger(x) = M^-1\Psi_D'^\dagger(x)M. \tag{53}$$

Consequently both the resulting equation of motion as well as the commutation relations differ from those of the original operators in a change of a sign of the factor e.

3. Consequences of the Furry picture

In the following we restrict ourselves to a class of fields where the definition of a stable vacuum is possible [4].

3.1. Vacuum polarization

Usually Feynman diagrams with self-closed electron lines, the tadpole diagrams, vanish as a consequence of Lorentz invariance. However, in the Furry picture the external field influences the vacuum: the vacuum expectation value of the current $j^\mu = \Psi^\dagger \gamma^\mu \Psi$ does not necessarily vanish any more. Therefore these kind of diagrams have to be included.

To make this plausible we split up the wave functions in states with positive and negative energy states:

$$\Psi_F^{(+)} = \sum_s c_s \Psi_{F,s}, \quad \text{and} \quad \Psi_F^{(-)} = \sum_\sigma c_\sigma \Psi_{F,\sigma}, \tag{54}$$

$$\Psi_F^{(\dagger,+)} = \sum_\sigma c_\sigma^* \Psi_{F,\sigma}^\dagger, \quad \text{and} \quad \Psi_F^{\dagger(-)} = \sum_s c_s^* \Psi_{F,s}^\dagger. \tag{55}$$

The vacuum expectation current can be expressed:

$$< j^\mu(x) > \sim \{\sum_s \Psi_s^\dagger \gamma^\mu \Psi_s - \sum_\sigma \Psi_\sigma^\dagger \gamma^\mu \Psi_\sigma\} \tag{56}$$

Eq.(56) is divergent. Using Pauli-Villar regularization [5], the equation can be transformed so that only in order e^2 logarithmically divergent but gauge invariant terms remain which can be removed via charge renormalization. Eq.(56) the whole vacuum polarization is contained up to the order e^2. In eq.(56)

3.2. Electron propagator

Since the external field influences the vacuum, space and time is no longer homogeneous. This inhomogenity causes the electron operator dependent on x and x' instead of only $(x - x')$.

The electron propagator can be defined:

$$G(x, x') = -i < 0|T\Psi_F(x)\Psi_F^\dagger(x')|0 > \tag{57}$$

4. Conclusions

The Furry representation is an intermediate picture between the Heisenberg and the Dirac representation. The crucial difference is that the eigenstates in the Furry picture are influenced by the external field whereas in the other representations one assumes a free-particle system as basis. This significant change has several consequences:

- the commutation relations for the wave functions in the Furry picture differ from those in the Dirac picture;
- the gauge transformation have to be accomplished by transformations including the external field;
- the charge conjugation is different for the wave functions in the Furry representation from those in the Dirac representation. As a consequence one gets different commutation relations for the wave functions and their charged-conjugated functions;
- as consequence of this lack of symmetry one gets contributions to self-closed diagrams in the vacuum polarization;
- another consequence of the lack of symmetry is that the Green functions of the electron operator is described by $G(x, x')$ instead of $G(x - x')$.

Acknowledgement
GMP would like to thank the Cockcroft Institute for the kind hospitality and the IPPP of Durham for the generous financial support. The author is also very thankful to the administrative secretaries Linda Wilkinson from the IPPP and Sue Waller from the CI who managed to cope with all eventualities.

References

[1] W. H. Furry, Phys. Rev. **81** (1951) 115.
[2] J. S. Schwinger, Phys. Rev. **74** (1948) 1439.
[3] J. S. Schwinger, Phys. Rev. **76** (1949) 790.
[4] W. H. Furry and J. R. Oppenheimer, Phys. Rev. **45** (1934) 245.
[5] W. Pauli and F. Villars, Rev. Mod. Phys. **21** (1949) 434.

Journal of Physics: Conference Series **198** (2009) 012003 doi:10.1088/1742-6596/198/1/012003

Recent development of quasiclassical operator method

V. N. Baier and V. M. Katkov
Budker Institute of Nuclear Physics, Novosibirsk, 630090, Russia
E-mail: V.N.Baier@inp.nsk.su

Abstract. The basis of the quasiclassical operator (QO) method is described. Application of the method for series of problems are discussed. These are processes in the superposition of plane wave and constant field, radiation in linear colliders, radiation in inhomogeneous fields, the new approach to the pair creation by a photon in a magnetic field, the theory of radiation in oriented crystals, radiation spectra taking into account the energy loss.

1. Introduction. Basis of QO method

The method was developed by authors in series of paper (BKp67), (BK67), (BK68), (BK69). The first paper was devoted to study of the radiative polarization of electrons moving in a magnetic field. In following years the method was applied for study of many problems including photon emission and pair creation in media, in oriented crystals, in linear colliders, in inhomogeneous external fields, in very strong fields interesting for astrophysics, radiation in periodic structures. List of the books and articles is given at the end of the paper.

There are two types of quantum effects at the radiation of high-energy particle in an external field (see Sec.2 in (BKS98)). The first one is associated with the quantization of particle motion in the field. For example, the commutator of the velocity components of relativistic particles in the magnetic field $\mathbf{H}$ (where the energy levels is $\varepsilon = \sqrt{m^2 + 2eH\hbar n} \gg m$) is

$$[v_i, v_k] = \frac{ie\hbar}{\varepsilon^2}\varepsilon_{ikj}H_j, \tag{1}$$

and the uncertainty relation for the velocity components reads

$$\Delta v_i \Delta v_k \sim \frac{e\hbar H}{\varepsilon^2} = \frac{H}{H_0\gamma^2} = \frac{\hbar\omega_0}{\varepsilon} \simeq \frac{1}{2n}, \quad H_0 = \frac{m^2}{e\hbar} = \left(\frac{m^2c^3}{\hbar e}\right) = 4.41 \cdot 10^{13}\mathrm{Oe}, \tag{2}$$

where we use units where $c = 1$, $\omega_0 = eH/\varepsilon$ is the Larmor frequency, $\gamma = \varepsilon/m$, so that with the energy rise the motion becomes increasingly classical.

The second type of quantum effects is associated with the recoil of particle at photon emission and is of the order $\hbar\omega/\varepsilon$. Already in the classical limit ($\hbar\omega \ll \varepsilon$) this type is principal, since $\omega \sim \omega_0\gamma^3$, and grows linearly with the energy rise.

The first order matrix element of the photon emission by a charged particle in the external field may be represented in the form

$$U_{fi} = \frac{ie}{2\pi\sqrt{\hbar\omega}}\int dt \int d^3r F^+_{fs'}(\mathbf{r})\exp(i\varepsilon_f t/\hbar)(e^*J)\exp[i(\omega t - \mathbf{kr})]\exp(-i\varepsilon_i t/\hbar)F_{is}(\mathbf{r}), \tag{3}$$

where $F_{is}(\mathbf{r})$ is the solution of the wave equation in the given field with the energy ε_i and in the spin state s, e^μ is the photon polarization vector, $k^\mu(\omega, \mathbf{k})$ is the photon 4-momentum, J^μ is the current vector.

For the states with $n \gg 1$ the following approximation may be made (see Eqs.(2.3)-(2.5) in (BKS98))

$$\exp(-i\varepsilon_i t/\hbar) F_{is}(\mathbf{r}) = \Psi_s(\mathbf{P}) \exp(-i\mathcal{H}t/\hbar)|i>, \quad P^\mu = i\hbar\partial^\mu - eA^\mu, \tag{4}$$

where $\Psi_s(\mathbf{P})$ is the operator form of the particle wave function in the spin state s in the given field. This form may be obtained from the free wave function via substitution of the variables by the operators: $\mathbf{p} \to \mathbf{P}$, $\varepsilon \to \mathcal{H} = \sqrt{\mathbf{P}^2 + m^2}$. In the coordinate representation $|i>$ is the solution of the Klein-Gordon equation in the given field.

Substituting Eq.(4) into Eq.(3) and taking into account that the Schrödinger operators, standing between the exponential factors $\exp(\pm i\mathcal{H}t/\hbar)$, convert into the time-dependent Heisenberg operators, we obtain the following formula for the matrix element Eq.(3) (see Eqs.(2.6)-(2.11) in (BKS98)):

$$U_{fi} = \langle f|M|i\rangle\,, \tag{5}$$

where

$$M = \frac{ie}{2\pi\sqrt{\hbar\omega}} \int dt \Psi^+_{s'}(\mathbf{p}) \{(e^* J), \exp[i(\omega t - \mathbf{k}\mathbf{r}(t))]\} \Psi_s(\mathbf{p}); \tag{6}$$

here $\mathbf{p}(t), J^\mu(t), \mathbf{r}(t)$ are the Heisenberg operators of the particle momentum, current and coordinates respectively, the brackets $\{,\}$ denote the symmetrized product of operators. Note, that $\Psi_s(\mathbf{p}) \exp(-i\mathcal{H}t/\hbar)\ |i>$ is the operator solution of the wave equation for the arbitrary form of the initial state $|i>$

We are interested in the transition probability with the photon emission summed over final particle states. Using the condition of the completeness of the states $\sum_f |f><f| = 1$ we obtain for the radiation probability

$$dw_\gamma =< i|M^+ M|i > d^3k. \tag{7}$$

The next step of calculation is series of operator transformations ("disentanglement") in Eqs.(7), (6). As a result one obtains (see Eqs.(2.12)-(2.28) in (BKS98))

$$M = \frac{e}{2\pi\sqrt{\hbar\omega}} \int_{-\infty}^{\infty} R(t) \exp\left[i\frac{kx(t)}{\varepsilon - \hbar\omega}\right] dt = \frac{e}{2\pi\sqrt{\hbar\omega}} \int_{-\infty}^{\infty} R(t) \exp\left[i\int_0^t \frac{kp(t')}{\varepsilon - \hbar\omega}dt'\right] dt, \tag{8}$$

here $R(t) = R(\mathbf{p}(t))$, $R(\mathbf{p})$ is the matrix element for the free particles depending on the particle spin, $kp = \omega\mathcal{H} - \mathbf{kp}$, $|i>$ is the state vector of the initial particle at the time $t = 0$, and $\mathbf{p}(t)$ is the operator of momentum in the Heisenberg picture.

The very important result of the method is that the recoil at radiation is incorporated into the theory in the *universal* form for any external field. There are two essentially different cases in application of the quasiclassical operator (QO) method. In the first case moving and scattering in a potential can be considered in *classical* terms: phase shifts are large, there is a correspondence between the impact parameter and the momentum transfer. Therefore it is possible to use the version of the QO method where one can substitute classical variables instead of operators in the expression Eq.(7) for the probability of process (see (BKp67),(BK68), (BKF72), (BLP82)). In the second case, the process of scattering can't be described in classical terms. However, for the processes where large angular momentum contributes one can use the *quasiclassical* approximation for description of scattering including the situation where the phase shifts are small . For example, this formulation of the method must be applied for consideration of radiation from ultrarelativistic particles at scattering on atoms in a media.

1.1. Macroscopic case (BK67), (BK68)

For the particle with spin 1/2 the function $R(t)$ in Eq.(8) written in the two-component form is (see Eqs.(2.37)-(2.47) and Eqs.(4.1)-(4.4) in (BKS98), below we employ units such that $\hbar = 1$)

$$R(t) = \varphi_{s'}^{+}(A + i\boldsymbol{\sigma}\mathbf{B})\varphi_s, \quad A = \frac{1}{2}\left(1 + \frac{\varepsilon}{\varepsilon'}\right)\mathbf{ev}, \quad \mathbf{B} = \frac{\omega}{2\varepsilon'}\mathbf{e}\times\mathbf{b}, \quad \mathbf{b} = \mathbf{n} - \mathbf{v} + \frac{\mathbf{n}}{\gamma}, \tag{9}$$

where $\mathbf{v} = \mathbf{v}(t)$ is the particle velocity, $\varepsilon' = \varepsilon - \omega$, $\gamma = \varepsilon/m$.

Substituting Eq.(8) into Eq.(7) we get $(t = (t_1 + t_2)/2,\ \tau = t_2 - t_1)$

$$dW \equiv \frac{dw_\gamma}{dt} = \frac{\alpha}{(2\pi)^2}\frac{d^3k}{\omega}\int R^*\left(t + \frac{\tau}{2}\right)R\left(t - \frac{\tau}{2}\right)$$
$$\times \exp\left\{-i\frac{\varepsilon}{\varepsilon'}\left[kx\left(t + \frac{\tau}{2}\right) - kx\left(t - \frac{\tau}{2}\right)\right]\right\}d\tau. \tag{10}$$

In the magnetic field, as in the classical theory, the radiation takes place from a small part of the trajectory on which the particle turns on the angle $\sim 1/\gamma$. This means that one can expand the integrand in Eq.(10) on τ powers. For the spectral distribution of the radiation intensity, for example, one obtains (see Eq.(4.26) in (BKS98))

$$\frac{dI}{du} = \frac{\alpha m^2}{\pi\sqrt{3}}\frac{u}{(1+u)^4}\left[(1 + (1+u)^2)K_{2/3}(\lambda) - (1+u)\int_\lambda^\infty K_{1/3}(z)dz\right], \tag{11}$$

where $K_\nu(z)$ is the MacDonald function,

$$u = \frac{\omega}{\varepsilon'}, \quad \lambda = \frac{2u}{3\chi}, \quad \chi = \frac{H}{H_0}\frac{\varepsilon}{m}, \tag{12}$$

here χ is the fundamental parameter. In the limit $\chi \ll 1$ one has the classical theory.

1.2. Microscopic case (BK69), Secs.16-18 (BKF73)

We consider the applicability of the QO method to the problem of radiation from a ultrarelativistic particle at potential scattering. In this process the emitted photon and the final electron are moving at small angles to the initial electron momentum and the large angular momenta $l \gg 1$ contribute. In this situation the quasiclassical scattering theory is applicable.

At very high energy of particle in a media, we can consider the case of complete screening, so that

$$a_s \ll \frac{1}{q_{min}} = \frac{2\varepsilon(\varepsilon - \omega)}{\omega m^2} \equiv l_f, \tag{13}$$

where a_s is the screening radius $(a_s \simeq 111Z^{-1/3}\lambda_c, \lambda_c = 1/m), Z$ is the charge of a nucleus, q_{min} is the minimal momentum transfer which is longitudinal (with respect to the momentum of the initial electron $\mathbf{p}$), l_f is the radiation formation length for the small angle scattering on an isolated atom. Note that in the frame of the QO method the radiation problem is solved for the case of arbitrary screening , see (BKF73). The impact parameters ϱ, contributing into the scattering cross section, are small comparing with the formation length $(\varrho \leq a_s \ll l_f)$ in the screened Coulomb potential. This means that the scattering of ultrarelativistic particles (the virtual electron is close to the mass shell) takes place independently from the radiation process (see, (BK69), (BKF73), (BLP82)). Thus, we can present the cross section of radiation as a product of the probability of photon emission with the momentum $\mathbf{k}$ at given momentum transfer $\mathbf{q}_\perp$ $(\mathbf{q}_\perp\mathbf{p} = 0)$, and the cross section of particle scattering $d\sigma(\mathbf{q}_\perp)$ with the same momentum transfer $\mathbf{q}_\perp$:

$$d\sigma_\gamma = W_\gamma(\mathbf{q}_\perp, \mathbf{k})d^3k\,d\sigma(\mathbf{q}_\perp), \tag{14}$$

Journal of Physics: Conference Series **198** (2009) 012003 doi:10.1088/1742-6596/198/1/012003

where the probability of photon emission is given in Eq.(7).

We show below that in the frame of the QO method the probability of radiation $W_\gamma(\mathbf{q}_\perp, \mathbf{k})$ is given by the trajectory of particle in "the form of an angle" in the momentum space

$$\mathbf{p}(t) = \vartheta(-t)\mathbf{p} + \vartheta(t)(\mathbf{p} + \mathbf{q}_\perp), \tag{15}$$

while the cross section $d\sigma(\mathbf{q}_\perp)$ should be taken in the *eikonal* form.

If the formation time of radiation is much longer than the characteristic time of scattering, one can present the dependence of the operator $\mathbf{p}(t)$ on time in Eq.(8) as

$$\begin{aligned}
&\mathbf{p}(t) = \vartheta(-t)\mathbf{p}(-\infty) + \vartheta(t)\mathbf{p}(\infty),\\
&\mathbf{p}_\perp(-\infty) \simeq \mathbf{p}_\perp + \int_{-\infty}^{z} \boldsymbol{\nabla}_\varrho V(\boldsymbol{\varrho}, z')dz', \quad \mathbf{p}_\perp(\infty) \simeq \mathbf{p}_\perp - \int_{z}^{\infty} \boldsymbol{\nabla}_\varrho V(\boldsymbol{\varrho}, z')dz'.
\end{aligned} \tag{16}$$

It should be pointed out that in the case when the scattering process is of nonclassical character the operators $\mathbf{p}(-\infty)$ and $\mathbf{p}(\infty)$ are noncommutative among themselves and, generally speaking, one can't neglect their commutator. But these approximate expressions have the classical form. Substituting the "trajectory" (16) in Eq.(8) we obtain

$$M = \frac{ie(\varepsilon - \omega)}{2\pi\sqrt{\omega}}\left[\frac{R(\mathbf{p}(\infty))}{kp(\infty)} - \frac{R(\mathbf{p}(-\infty))}{kp(-\infty)}\right]. \tag{17}$$

For the derivation of the differential cross section of bremsstrahlung in Eqs.(14), (7) it is necessary to insert the projection operator $|f><f|$ between the operators M^+ and M, and to take $|i>$ and $|f>$ states so that the initial state is the eigenvector of the operator $\mathbf{p}(-\infty)$ and the final state is the eigenvector of the operator $\mathbf{p}(\infty)$:

$$\begin{aligned}
&\mathbf{p}(-\infty)|i>= \mathbf{p}_i|i>, \quad |i>= \exp\left[-i\int_{-\infty}^{z} V(\boldsymbol{\varrho}, z')dz'\right]|\mathbf{p}_i>,\\
&\mathbf{p}(\infty)|f>= \mathbf{p}_f|f>, \quad |f>= \exp\left[i\int_{z}^{\infty} V(\boldsymbol{\varrho}, z')dz'\right]|\mathbf{p}_f>.
\end{aligned} \tag{18}$$

Using (18) we deduce for the matrix element of the operator M Eq.(17)

$$\begin{aligned}
&M_{fi} = \frac{ie(\varepsilon - \omega)}{2\pi\sqrt{\omega}}\left[\frac{R(\mathbf{p}_f)}{kp_f} - \frac{R(\mathbf{p}_i)}{kp_i}\right] < f|i>,\\
&< f|i>=< \mathbf{p}_f|\exp\left[-i\int_{-\infty}^{\infty} V(\boldsymbol{\varrho}, z)dz\right]|\mathbf{p}_i>= \int d^2\varrho \exp\left[i\mathbf{q}_\perp\boldsymbol{\varrho} + i\chi(\boldsymbol{\varrho})\right] 2\pi\delta(p_{f\|} - p_{i\|}),
\end{aligned} \tag{19}$$

where

$$\mathbf{q}_\perp = \mathbf{p}_{i\perp} - \mathbf{p}_{f\perp}, \quad \chi(\boldsymbol{\varrho}) = -\int_{-\infty}^{\infty} V(\boldsymbol{\varrho}, z)dz. \tag{20}$$

In centrally symmetric potential we have

$$\chi(\varrho) = -\int_{-\infty}^{\infty} V(\sqrt{\varrho^2 + z^2})dz. \tag{21}$$

We introduce now the notations $\mathbf{p}_i \equiv \mathbf{p}$, $\mathbf{p}_f = \mathbf{p}' + \mathbf{k}$ where $\mathbf{p}'$ is the momentum of electron after photon emission. Then, neglecting the terms of the order of $q_\|$ in the argument of the δ-function in Eq.(19), we have in the region $q_\perp \gg q_\|$, which contributes for the potential considered,

$$\delta(p_{f\|} - p_{i\|}) \simeq \delta(\varepsilon' + \omega - \varepsilon), \quad \varepsilon' = \sqrt{\mathbf{p}'^2 + m^2}. \tag{22}$$

Using this relation one can express the propagator kp_f through the propagator kp' (up to terms of the order $1/\gamma^2$)

$$kp_f = \omega\sqrt{(\mathbf{p}'+\mathbf{k})^2+m^2} - \mathbf{k}(\mathbf{p}'+\mathbf{k})$$
$$= \omega\sqrt{\varepsilon^2 - 2kp'} + kp' - \omega\varepsilon \simeq \omega\varepsilon\left(1-\frac{kp'}{\varepsilon^2}\right) + kp' - \omega\varepsilon = \frac{\varepsilon'}{\varepsilon}kp'. \tag{23}$$

Our final result for the differential probability of radiation in Eq.(14) is therefore

$$W_\gamma(\mathbf{q}_\perp,\mathbf{k}) = \frac{\alpha}{(2\pi)^2}\frac{1}{\omega}\left|\frac{\varepsilon R(\mathbf{p}'+\mathbf{k})}{kp'} - \frac{\varepsilon' R(\mathbf{p})}{kp}\right|^2. \tag{24}$$

Using the Moliere approximation of the atomic potential for the phase $\chi(\varrho)$ in Eq.(21) we have for the spectrum of bremsstrahlung in the case of complete screening

$$\frac{d\sigma}{d\omega} = \frac{4Z^2\alpha^3\varepsilon'}{m^2\varepsilon\omega}\left[\left(\frac{\varepsilon}{\varepsilon'}+\frac{\varepsilon'}{\varepsilon}-\frac{2}{3}\right)\left(\ln(183Z^{-1/3}) - f(Z\alpha)\right)+\frac{1}{9}\right], \tag{25}$$

where

$$f(\xi) = \xi^2\sum_{n=1}^{\infty}\frac{1}{n(n^2+\xi^2)}, \tag{26}$$

here the function $f(\xi)$ is the Coulomb correction. In the Coulomb field the term 1/9 in the square brackets should be omitted and $\ln(183Z^{-1/3})$ should be substituted by $\ln(1/\delta) - 1/2$, where $\delta = q_{min}/m = \omega m/2\varepsilon\varepsilon'$.

2. Processes in plane wave and constant field (BKS91)

Substituting Eqs.(8),(9) into Eq.(7) and performing the integration over the photon emission angle and the summation over the polarization of final particles, we obtain the spectral distribution of radiation in the convenient form, where all cancelations in the leading terms have already been carried out (see also Eqs.(2.37)-(2.47) in (BKS98)):

$$\frac{dw_\gamma}{d\omega} = \frac{i\alpha}{8\pi\gamma^2}\int\int\frac{dt d\tau}{\tau - i0}\left[4+\beta(\mathbf{\Delta}_1-\mathbf{\Delta}_2)^2\right]\exp\left\{-\frac{i\tau}{l_\omega}\left(1+\int_{t_1}^{t_2}\mathbf{\Delta}^2(t')dt'\right)\right\},$$
$$\beta = \left(\frac{\varepsilon}{\varepsilon'}+\frac{\varepsilon'}{\varepsilon}\right),\quad \mathbf{\Delta}_1 = \mathbf{\Delta}(t_1),\quad \mathbf{\Delta} = \mathbf{\Delta}(t) = \frac{1}{m}(\mathbf{p}(t) - \boldsymbol{\pi}),$$
$$\boldsymbol{\pi} = \frac{1}{\tau}\int_{t_1}^{t_2}\mathbf{p}(t)dt,\quad l_\omega = \frac{2\varepsilon\varepsilon'}{m^2\omega}, \tag{27}$$

where $\mathbf{p} = \varepsilon\mathbf{v}$ is the momentum of the particle. Obviously, the quantity $\mathbf{\Delta}(t)$ is unaffected by the substitution $\mathbf{p}(t) \to \mathbf{p}(t) + \mathbf{p}_0$, where $\mathbf{p}_0$ is the time-independent momentum. To pursue the analysis we need explicit expressions for the momentum $\mathbf{p}(t)$ and the vector $\mathbf{\Delta}(t)$ in the field under consideration. We will carry out calculations in the frame of reference in which the monochromatic plane wave, with the wave vector $q = q(q_0, \mathbf{q})$, is propagating in the direction $\mathbf{n} = \mathbf{q}/q_0$ opposite to the electron velocity. One can always find a relativistic frame of reference ($\gamma \gg 1$) in which the condition $q_0 \ll \varepsilon$ holds. This is necessary condition if we wish to treat the plane wave as classical. Solving the equation of particle motion in the electromagnetic field we get

$$\frac{\mathbf{p}_\perp}{m} = \mathbf{\Omega}t + \boldsymbol{\xi}(t),\quad \boldsymbol{\xi}(t) = \boldsymbol{\xi}_2\sin(\nu t+\varphi_0) + \boldsymbol{\xi}_1\cos(\nu t+\varphi_0),$$
$$\mathbf{\Omega} = \frac{e}{m}\mathbf{F}_\perp,\quad \mathbf{F}_\perp = \mathbf{E} - \mathbf{n}(\mathbf{n}\mathbf{E}) + \mathbf{H}\times\mathbf{n},\quad \nu = 2q_0,\quad \boldsymbol{\xi}\mathbf{n} = 0. \tag{28}$$

Here **E** and **H** are the electric and magnetic fields, both independent from time, and the orthogonal vectors $\boldsymbol{\xi}_{1,2}$ characterize the intensity $\xi_0^2 = (\xi_1^2 + \xi_2^2)/2$ and the polarization of the wave. The corresponding Stokes parameters are

$$\lambda_3 = \frac{\xi_1^2 - \xi_2^2}{\xi_1^2 + \xi_2^2}, \quad \lambda_2 = \frac{(\boldsymbol{\xi}_1 \times \boldsymbol{\xi}_2)\mathbf{n}}{\xi_0^2}. \tag{29}$$

Performing calculation we obtain the spectral probability of photon emission per unit time

$$\frac{d^2 w_\gamma(t)}{dtdy} = \frac{dW_\gamma}{dy} = \frac{i\alpha m^2}{2\pi\varepsilon} \int\limits_{-\infty}^{\infty} \frac{d\tau}{\tau - i0} \left[1 + \left(\frac{\varepsilon}{\varepsilon'} + \frac{\varepsilon'}{\varepsilon}\right) m_0\right] \exp(-iu\tau\Phi(\varphi, \tau)), \tag{30}$$

where

$$y = \frac{\omega}{\varepsilon}, \quad u = \frac{y}{1-y}, \quad m_0 = \chi^2\tau^2 + 2\boldsymbol{\chi\eta}(\varphi)\tau \sin\frac{s\tau}{2} + \xi_0^2(1 - \lambda_3 \cos 2\varphi)\sin^2\frac{s\tau}{2},$$
$$\Phi = \frac{1}{3}\chi^2\tau^2 + \frac{8}{s^2\tau}\boldsymbol{\chi\eta}(\varphi)\left(\sin\frac{s\tau}{2} - \frac{s\tau}{2}\cos\frac{s\tau}{2}\right) +$$
$$\xi_0^2\left[1 + \frac{2}{s^2\tau^2}(\cos s\tau - 1) + \frac{\lambda_3}{s\tau}\cos 2\varphi\left(\sin s\tau + \frac{2}{s\tau}(\cos s\tau - 1)\right)\right] + 1,$$
$$\boldsymbol{\eta}(\varphi) = \boldsymbol{\xi}_2 \cos\varphi - \boldsymbol{\xi}_1 \sin\varphi, \quad \varphi = \nu t + \varphi_0, \quad \tau \to l_0\tau, \ l_0 = \frac{\varepsilon}{m^2},$$
$$\boldsymbol{\Omega} l_0 = \frac{e}{m}\mathbf{F}_\perp l_0 = \boldsymbol{\chi}, \quad 2\nu l_0 = \frac{4q_0\varepsilon}{m^2} \simeq \frac{2qp}{m^2} = s. \tag{31}$$

If the plane wave is absent ($\boldsymbol{\xi}_1 = \boldsymbol{\xi}_2 = 0$) Eq.(31) turns into the spectral probability in constant field in the quasiclassical approximation. In absence of constant field ($\boldsymbol{\chi} = 0$) Eq.(31) turns into the *exact* spectral probability in the monochromatic plane wave. In the case $\xi_0 \ll 1$ one can expand the exponent in Eq.(31) over $\xi(\eta)$ and retain the terms $\propto \xi_0^2$:

$$dW_\gamma = dW_\gamma^F + \xi_0^2 dW_\gamma^\xi, \tag{32}$$

where dW_γ^F is the spectral probability in constant field and dW_γ^ξ is connected with the cross section of the Compton effect

$$d\sigma_c = \frac{8\pi\alpha\varepsilon}{m^2 s} dW_\gamma^\xi. \tag{33}$$

The cross section $d\sigma_c$ describes the photon scattering on electron in the presence of an external field. A number of specific examples of the Compton effect in a static field is given in (BKS91).

3. Radiation in linear colliders (BKS90, BK04)

The particle interaction at beam-beam collision in linear colliders occurs in an electromagnetic field provided by the beams. The magnetic bremsstrahlung mechanism dominates and its characteristics are determined by the value of the quantum parameter $\chi(t)$ dependent on the strength of the incoming beam field at the moment t (the constant field limit) $\chi = \gamma F/H_0$, where $F = |\mathbf{F}|$, $\mathbf{F} = \mathbf{E}_\perp + \mathbf{v} \times \mathbf{H}$, $\mathbf{E}$ and $\mathbf{H}$ are the electric and magnetic fields in the laboratory frame, $\mathbf{E}_\perp = \mathbf{E} - \mathbf{v}(\mathbf{vE})$.

The photon radiation length in an external field is

$$l_c(\chi, u) = \lambda_c \frac{H_0}{F}\left(1 + \frac{\chi}{u}\right)^{1/3} = \frac{\lambda_c\gamma}{\chi}\left(1 + \frac{\chi}{u}\right)^{1/3}. \tag{34}$$

Journal of Physics: Conference Series **198** (2009) 012003 doi:10.1088/1742-6596/198/1/012003

The field of the incoming beam changes very slightly along the formation length l_c, if the condition $l_c \ll \sigma_z$ if satisfied, providing a high accuracy of the magnetic bremsstrahlung approximation.

In the general case, when both polarization of electrons and photons is taken into account, the spectral probability of radiation per unit time has the form

$$\frac{dw_\gamma}{dt} \equiv dW_\gamma(t) = \frac{\alpha}{2\sqrt{3}\pi\gamma^2}\Phi_\gamma^\zeta(1+(\boldsymbol{\lambda}\boldsymbol{\xi}))d\omega; \quad \Phi_\gamma^\zeta = \Phi_\gamma + \frac{\omega}{\varepsilon}(\boldsymbol{\zeta}\mathbf{h})K_{1/3}(z),$$
$$\Phi_\gamma(t) = \beta K_{2/3}(z) - \int_z^\infty K_{1/3}(y)dy, \quad \beta = \frac{\varepsilon}{\varepsilon'} + \frac{\varepsilon'}{\varepsilon}, \tag{35}$$

where $z = 2u/3\chi(t)$, $\boldsymbol{\lambda}(\lambda_1, \lambda_2, \lambda_3)$ are the Stokes parameters of emitted photons for the following choice of axes: $\mathbf{e}_1 = (\mathbf{v}\times\mathbf{h})$, $\mathbf{h} = \mathbf{F}^*/F$, $\mathbf{e}_2 = \mathbf{h}$, $\mathbf{F}^* = e/|e|[\mathbf{H}_\perp + (\mathbf{E}\times\mathbf{v})]$, $\boldsymbol{\zeta}$ is the spin vector of the initial electron in its rest frame. The vector $\boldsymbol{\xi}$ determines the mean photon polarization and its components are given by the following expressions:

$$\xi_1 = \frac{\omega(\boldsymbol{\zeta}\mathbf{v}\mathbf{h})}{\varepsilon'\Phi_\gamma^\zeta}K_{1/3}(z), \quad \xi_2 = \frac{(\boldsymbol{\zeta}\mathbf{v})}{\Phi_\gamma^\zeta}\left[\left(\frac{\varepsilon}{\varepsilon'} - \frac{\varepsilon'}{\varepsilon}\right)K_{2/3}(z) - \frac{\omega}{\varepsilon}\int_z^\infty K_{1/3}(y)dy\right],$$
$$\xi_3 = \frac{1}{\Phi_\gamma^\zeta}\left[K_{2/3}(z) + \frac{\omega}{\varepsilon'}(\boldsymbol{\zeta}\mathbf{h})K_{1/3}(z)\right], \tag{36}$$

here $(\boldsymbol{\zeta}\mathbf{v}\mathbf{h}) = \boldsymbol{\zeta}(\mathbf{v}\times\mathbf{h})$.

Below we consider radiation from unpolarized electrons. The spectral probability of radiation is (35)

$$\frac{dw_\gamma}{d\omega} = \frac{\alpha}{\pi\gamma^2\sqrt{3}}\int_{-\infty}^{\infty}\Phi_\gamma(t)dt. \tag{37}$$

For the Gaussian beams

$$\chi(t) = \chi_0(x,y)\exp(-2t^2/\sigma_z^2), \tag{38}$$

here the function $\chi_0(x,y)$ depends on transverse coordinates.

It turns out that for the Gaussian beams the integration of the spectral probability over time can be carried out in a general form:

$$\frac{dw_\gamma}{du} = \frac{\alpha m\sigma_z}{\pi\gamma\sqrt{6}}\frac{1}{(1+u)^2}\left[\left(1+u+\frac{1}{1+u}\right)\int_1^\infty K_{2/3}(ay)\frac{dy}{y\sqrt{\ln y}}\right.$$
$$\left.-2a\int_1^\infty K_{1/3}(ay)\sqrt{\ln y}\,dy\right], \tag{39}$$

where $a = 2u/3\chi_0$. In the case when $\chi_0 \ll 1$ the main contribution into integral (39) gives the region $y = 1+\xi, \xi \ll 1$. Taking the integrals over ξ we obtain

$$\frac{dw_\gamma^{CF}}{du} \simeq \frac{\sqrt{3}\alpha m\sigma_z}{4\gamma}\frac{1+u+u^2}{u(1+u)^3}\chi_0\exp\left(-\frac{2u}{3\chi_0}\right). \tag{40}$$

For the round beams the integration over transverse coordinates is performed with the density

$$n_\perp(\boldsymbol{\varrho}) = \frac{1}{2\pi\sigma_\perp^2}\exp\left(-\frac{\varrho^2}{2\sigma_\perp^2}\right). \tag{41}$$

The parameter $\chi_0(\varrho)$ we present in the form

$$\chi_0(\varrho) = \chi_m \frac{f(x)}{f_0}, \quad x = \frac{\varrho}{\sigma_\perp}, \quad f(x) = \frac{1}{x}\left(1 - \exp(-x^2/2)\right),$$
$$\chi_{rd} = 0.720\alpha N\gamma \frac{\lambda_c^2}{\sigma_z \sigma_\perp}, \quad f'(x_0) = 0, \; f_0 = f(x_0) = 0.451256, \tag{42}$$

where N is the number of electrons in the bunch. The result of calculation is shown in Fig.1.

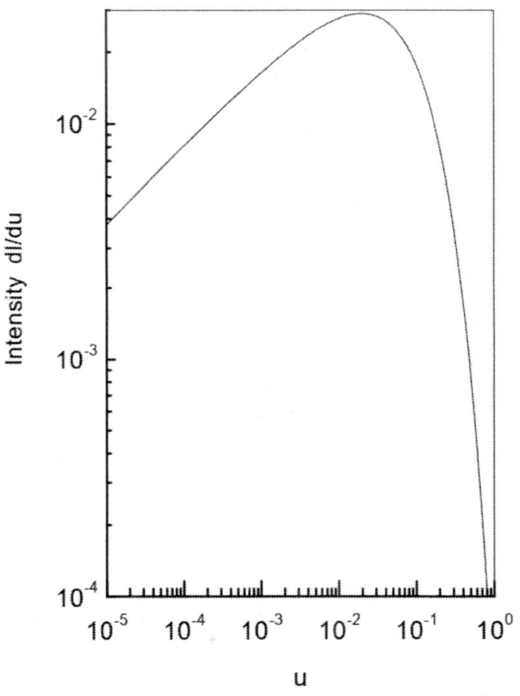

Figure 1. The spectral intensity of radiation from the round beams in units $\alpha m^2\sigma_z$ for χ_{rd}=0.13 calculated according to Eqs.(39),(42).

The Laplace integration of Eq.(40) gives for the radiation intensity $dI/du = \varepsilon u/(1+u)dW/du$

$$\frac{dI_{as}}{du} \simeq \alpha m^2 \sigma_z \frac{3}{4}\sqrt{\frac{\pi}{|f_0''|}} \frac{1+u+u^2}{\sqrt{u}(1+u)^4} f_0^{3/2} \chi_{rd}^{3/2} \exp\left(-\frac{2u}{3\chi_m}\right), \tag{43}$$

where $f_0'' = f''(x_0) = -0.271678$.

For the flat beams ($\sigma_x \gg \sigma_y$) the parameter $\boldsymbol{\chi}_0(\boldsymbol{\varrho})$ takes the form

$$\boldsymbol{\chi}_0 = \chi_m \exp\left(-\frac{x^2}{2\sigma_x^2}\right)\left[\mathbf{e}_y \mathrm{erf}\left(\frac{y}{\sqrt{2}\sigma_y}\right) - i\mathbf{e}_x \mathrm{erf}\left(i\frac{x}{\sqrt{2}\sigma_x}\right)\right], \quad \chi_m = \frac{2N\alpha\gamma\lambda_c^2}{\sigma_z \sigma_x}, \tag{44}$$

here $\mathrm{erf}(z) = 2/\sqrt{\pi}\int_0^z \exp(-t^2)dt$, $\mathbf{e}_x$ and $\mathbf{e}_y$ are the unit vectors along the corresponding axes.

To calculate the asymptotic of radiation intensity for the case $\chi_0 \ll 1$ one has to substitute

$$\chi_0 = |\boldsymbol{\chi}_0| = \chi_m \exp\left(-\frac{x^2}{2\sigma_x^2}\right)\left[\left(\text{erf}\left(\frac{y}{\sqrt{2}\sigma_y}\right)\right)^2 + \left(-i\text{erf}\left(i\frac{x}{\sqrt{2}\sigma_x}\right)\right)^2\right]^{1/2} \tag{45}$$

into Eq.(40) and take integrals over the transverse coordinates x, y with the weight

$$n_\perp(x,y) = \frac{1}{2\pi\sigma_x\sigma_y}\exp\left(-\frac{x^2}{2\sigma_x^2} - \frac{y^2}{2\sigma_y^2}\right). \tag{46}$$

Integral over x can be taken using the Laplace method, while at integration over y it is convenient to introduce the variable

$$\eta = \frac{2}{\sqrt{\pi}}\int_w^\infty \exp(-t^2)dt, \quad w = \frac{y}{\sqrt{2}\sigma_y}. \tag{47}$$

As a result we obtain for the radiation intensity in the case of flat beams

$$\frac{dI_{fl}}{du} = \frac{9}{8\sqrt{2\left(1-2/\pi\right)}}\alpha m^2\sigma_z\chi_m^{5/2}\frac{1+u+u^2}{u^{3/2}(1+u)^4}\exp\left(-\frac{2u}{3\chi_m}\right). \tag{48}$$

4. Inhomogeneous fields (BKS89)

If the field varies slightly on the photon formation length, the vector $\boldsymbol{\Delta}(t_2)$ in Eq.(27) as well as the exponential factor can be expanded in powers of $t_2 - t_1 = \tau$, with the required number of expansion terms.The first terms of the expansion which incorporates the particle acceleration, give the constant field limit, while remaining terms are the correction to this approximation. As the result the spectral intensity of radiation can be presented in the form

$$\frac{dI}{d\omega} = \frac{dI_0}{d\omega} + \frac{dI_c}{d\omega}, \quad \frac{dI_0}{d\omega} = \frac{\alpha m^2\omega}{\sqrt{3}\pi\varepsilon^2}\left[\beta K_{2/3}(z) - \int\limits_z^\infty K_{1/3}(y)dy\right],$$
$$\frac{dI_c}{d\omega} = \frac{\alpha m^2\omega}{\sqrt{3}\pi\varepsilon^2}\Bigg\{-\frac{1}{3\mathbf{b}^4}\Bigg[(\mathbf{b}(\mathbf{V}\boldsymbol{\nabla})^2\mathbf{b})\beta\left[K_{2/3}(z) - \frac{2}{3z}K_{1/3}(z)\right]$$
$$-\frac{1}{10}\left[((\mathbf{V}\boldsymbol{\nabla})\mathbf{b})^2 + 3(\mathbf{b}(\mathbf{V}\boldsymbol{\nabla})^2\mathbf{b})\right]$$
$$\times\left[zK_{1/3}(z) - \frac{4}{3}K_{2/3}(z) + \beta\left(4K_{2/3}(z) - \left(z + \frac{16}{9z}\right)K_{1/3}(z)\right)\right]\Bigg]\Bigg\}, \tag{49}$$

where β is defined in Eq.(35),

$$z = \frac{2m^2\omega}{3\varepsilon\varepsilon'|\mathbf{b}|}, \quad \mathbf{b} = \frac{e\mathbf{F}}{m}, \tag{50}$$

the vector $\mathbf{V}$ is the difference between the velocity of particle and the opposite beam (in the case when the beam shape is not changed during the collision). Here $dI_0/d\omega$ is the intensity spectrum in magnetic bremsstrahlung limit, $dI_c/d\omega$ is the gradient correction.

5. Pair creation by a photon in an external field (BKS98, BK07)

The probability of pair creation by a photon in the frame of QO method may be obtained from the probability of radiation from electron $dW = dI/\omega$ Eq.(11) using the QED crossing symmetry:$p \to -p, k \to -k$, the final electron momentum $p' \to p'$, $d^3k \to -d^3p$. This law is exact for the differential probabilities. However, since in these processes all the final particles are moving within the angle $\sim 1/\gamma$ in the forward direction, the crossing symmetry relation are fulfilled within the relativistic accuracy also for the probabilities integrated over angles, i.e. for the spectral distributions. Substituting in Eq.(11) $\varepsilon \to -\varepsilon$, $\omega \to -\omega$, $\omega^2 d\omega \to -\varepsilon^2 d\varepsilon$ one obtains for the spectral distribution of created particle (see Eq.(3.50) in (BKS98))

$$dW = \frac{\alpha m^2}{\pi\sqrt{3}}\frac{dx}{\omega}\left[\left(\frac{x}{1-x}+\frac{1-x}{x}\right)K_{2/3}(\xi)+\int\limits_{\xi}^{\infty}K_{1/3}(z)dz\right], \tag{51}$$

where

$$x=\frac{\varepsilon}{\omega},\quad \xi=\frac{2\omega}{3\kappa x(1-x)},\quad \kappa=\frac{H\omega}{H_0 m}, \tag{52}$$

ω is the energy of the initial photon, ε is the energy of created positron and $\omega-\varepsilon$ is the energy of created electron.

Integrating over energy of created particle one obtains the total probability of pair creation (see Eq.(3.56) in (BKS98))

$$W=\frac{\alpha m^2}{3\sqrt{3}\pi\omega}\int\limits_{-1}^{1}\frac{9-v^2}{1-v^2}K_{2/3}(\xi)dv, \tag{53}$$

where $\xi = 8/[3\kappa(1-v^2)]$ (the substitution was made $x(1-x)=(1-v^2)/4$).

In the limit $\kappa \ll 1$ the variable $\xi \gg 1$ and one can substitute into the integral Eq.(53) the asymptotic $K_\nu(z)=\sqrt{\pi/2z}\exp(-z)$. After integration one obtains in this limit

$$W^{(QO)}=\frac{3\sqrt{3}\alpha m^2\kappa}{16\sqrt{2}\omega}\exp\left(-\frac{8}{3\kappa}\right). \tag{54}$$

So at $\kappa \ll 1$ the probability of pair creation is exponentially small. Such situation is characteristic for all processes with the finite discontinuity of the square of system four momentum (the invariant mass of system):$\Delta_p^2=(p+p')^2-k^2=(p+p')^2$, $\Delta^2_{p\ min}=4m^2$.

In the case when the magnetic field is weak comparing with critical one and for high energy photons

$$\mu\equiv\frac{H}{H_0}\ll 1,\quad r\equiv\frac{k_\perp^2}{4m^2}\gg 1,\quad \mathbf{k}_\perp\mathbf{H}=0,\quad \kappa=2\sqrt{r}\mu\ll 1, \tag{55}$$

the standard QO method given above is applicable if $\kappa \gg 1/r$.

At $r \leq \mu^{-2/3}$ this method becomes inapplicable. In this region one can use the process probability found in Appendix B of (BK07), which is valid if $\mu \ll r-1 \ll \mu^{-2}$. Using this result we get for the probability of pair creation by unpolarized photon

$$W^{(th)}=\frac{\alpha m^2\mu}{4\omega}\frac{3r-1}{\sqrt{r(r-1)l(r)\beta(r)}}\exp\left(-\frac{\beta(r)}{\mu}\right), \tag{56}$$

where

$$l(r)=\ln\frac{\sqrt{r}+1}{\sqrt{r}-1},\quad \beta(r)=2\sqrt{r}-(r-1)l(r). \tag{57}$$

For large $r \gg 1$, limited by the condition $r \ll \mu^{-2}$ ($\kappa \ll 1$), we have

$$W^{(th)}\simeq\frac{3\alpha m^2\mu}{8\omega}\sqrt{\frac{3r}{2}}\exp\left(-\frac{4}{3\mu\sqrt{r}}-\frac{4}{15\mu r^{3/2}}\right)=W^{(QO)}\exp\left(-\frac{4}{15\mu r^{3/2}}\right). \tag{58}$$

At $\kappa^3 \gg \mu^2$ Eq.(58) agrees with Eq.(54). The last expression has essentially wider region of applicability than pure $W^{(QO)}$. It is seen that the probability of pair creation depends separately on field (parameter μ) and photon energy (parameter r), while in standard QO method this probability depends on combination $\sqrt{r}\mu$ only. The ratio $W^{(th)}/W^{(QO)}$ is shown in Fig.2. One can see that with μ decreasing the photon energy interval, in which Eq.(56) should be used, is significantly extended.

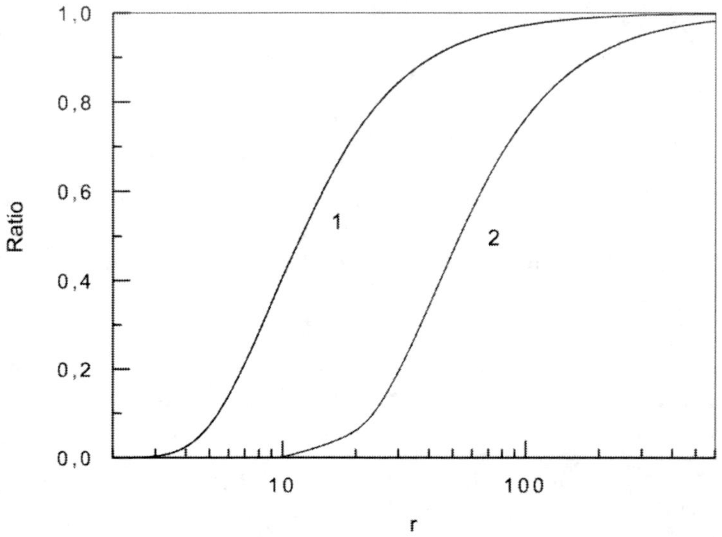

Figure 2. The ratio of the pair creation probability $W^{(th)}$ Eq.(56) to the standard $W^{(QO)}$ Eq.(54) vs the photon energy parameter r for $\mu = 0.01$ (curve 1) and for $\mu = 0.001$ (curve 2).

At $\kappa \geq 1$ Eq.(56) becomes inapplicable. However in this case QO is valid, see Eq.(53).

6. General theory of radiation in oriented crystal (BK05, BK06)

Here we consider radiation from electron in an oriented single crystal for the case when the angle of incidence (angle between the electron momentum and axis) $\vartheta_0 \ll V_0/m$, V_0 is the scale of the axis potential (see below). In this case the distance of an electron from axis $\boldsymbol{\varrho}$ as well as the transverse field of the axis can be considered as constant at the formation length. For an axial orientation of crystal the ratio of the atom density $n(\varrho)$ in the vicinity of an axis to the mean atom density n_a is

$$\frac{n(x)}{n_a} = \xi(x) = \frac{x_0}{\eta_1} e^{-x/\eta_1}, \quad \varepsilon_0 = \frac{\varepsilon_e}{\xi(0)}, \tag{59}$$

where

$$x_0 = \frac{1}{\pi d n_a a_s^2}, \quad \eta_1 = \frac{2u_1^2}{a_s^2}, \quad x = \frac{\varrho^2}{a_s^2}. \tag{60}$$

Here ϱ is the distance from axis, u_1 is the amplitude of thermal vibration, d is the mean distance between atoms forming the axis, a_s is the effective screening radius of the axis potential

$$U(x) = V_0 \left[\ln\left(1 + \frac{1}{x + \eta}\right) - \ln\left(1 + \frac{1}{x_0 + \eta}\right) \right]. \tag{61}$$

Journal of Physics: Conference Series **198** (2009) 012003 doi:10.1088/1742-6596/198/1/012003

The local value of parameters $\chi(x)$, see Eq.(12), which determines the radiation probability in the field Eq.(61) is

$$\chi(x) = -\frac{dU(\varrho)}{d\varrho}\frac{\varepsilon}{m^3} = \chi_s f_a, \quad f_a = \frac{2\sqrt{x}}{(x+\eta)(x+\eta+1)}, \quad \chi_s = \frac{V_0\varepsilon}{m^3 a_s} \equiv \frac{\varepsilon}{\varepsilon_s}. \tag{62}$$

The particular calculation below will be done for the tungsten and germanium crystals studied experimentally. The relevant parameters are given in Table 1.

TABLE 1 Parameters of radiation (pair creation) process in the tungsten (the axis < 111 >) and germanium (the axis < 110 >) crystals for different temperatures T, the energies ε and ω are in GeV

Crystal	T(K)	V_0(eV)	x_0	η_1	η	ε_0	ε_t	$\varepsilon_s(\omega_s)$	$\varepsilon_m(\omega_m)$	h
W	293	413	39.7	0.108	0.115	7.43	0.76	34.8	14.35	0.348
W	100	355	35.7	0.0401	0.0313	3.06	0.35	43.1	8.10	0.612
Ge	100	114.5	19.8	0.064	0.0633	59	0.85	179	51	0.459

Here ε_t is an estimate of electron energy when the intensity of coherent radiation is equal to the intensity of incoherent one. The maximal value of the parameter $\chi(x)$ is $\chi_m = \chi(x_m) = \varepsilon/\varepsilon_m$.

The spectral intensity of radiation can be presented in the form

$$\begin{aligned}
&dI(\varepsilon, y) = \frac{\alpha m^2}{2\pi}\frac{ydy}{1-y}\int\limits_0^{x_0}\frac{dx}{x_0}G(x,y), \quad G(x,y) = \int\limits_0^{\infty}F(x,y,t)dt - r_3\frac{\pi}{4}, \\
&F(x,y,t) = \mathrm{Im}\left\{e^{\varphi_1(t)}\left[r_2\nu_0^2(1+ib)\varphi_2(t) + r_3\varphi_3(t)\right]\right\}, \quad b = \frac{4\chi^2(x)}{u^2\nu_0^2}, \\
&y = \frac{\omega}{\varepsilon}, \quad u = \frac{y}{1-y}, \quad \varphi_1(t) = (i-1)t + b(1+i)(f_2(t) - t), \\
&\varphi_2(t) = \frac{\sqrt{2}}{\nu_0}\tanh\frac{\nu_0 t}{\sqrt{2}}, \quad \varphi_3(t) = \frac{\sqrt{2}\nu_0}{\sinh(\sqrt{2}\nu_0 t)},
\end{aligned} \tag{63}$$

where

$$\begin{aligned}
&r_2 = 1 + (1-y)^2, \quad r_3 = 2(1-y), \quad \nu_0^2 = \frac{1-y}{y}\frac{\varepsilon}{\varepsilon_c(x)}, \quad \varepsilon_c(x) = \frac{\varepsilon_e(n_a)}{\xi(x)g} = \frac{\varepsilon_0}{g}e^{x/\eta_1}, \\
&\varepsilon_e = \frac{m}{16\pi Z^2\alpha^2\lambda_c^3 n_a L_0}, \quad L_0 = \ln(183Z^{-1/3}) - f(Z\alpha), \\
&h(z) = -\frac{1}{2}\left[1 + (1+z)e^z\mathrm{Ei}(-z)\right], \quad g = 1 + \frac{1}{L_0}\left[\frac{1}{18} - h\left(\frac{u_1^2}{a^2}\right)\right].
\end{aligned} \tag{64}$$

The found spectral intensity of radiation contains a very rich information. The intensity of coherent radiation $I(\varepsilon) = \int I^{coh}(\varepsilon, y)dy$ is the first term ($\nu_0^2 = 0$) of the decomposition of Eq.(63) over ν_0^2

$$I^{coh}(\varepsilon) = \int\limits_0^{x_0} I(\chi)\frac{dx}{x_0}. \tag{65}$$

Here $I(\chi)$ is the radiation intensity in constant field (magnetic bremsstrahlung limit). It is convenient to use the following representation for $I(\chi)$

$$I(\chi) = i\frac{\alpha m^2}{2\pi}\int\limits_{\lambda-i\infty}^{\lambda+i\infty}\left(\frac{\chi^2}{3}\right)^s\Gamma(1-s)\Gamma(3s-1)(2s-1)(s^2-s+2)\frac{ds}{\cos\pi s}, \quad \frac{1}{3} < \lambda < 1. \tag{66}$$

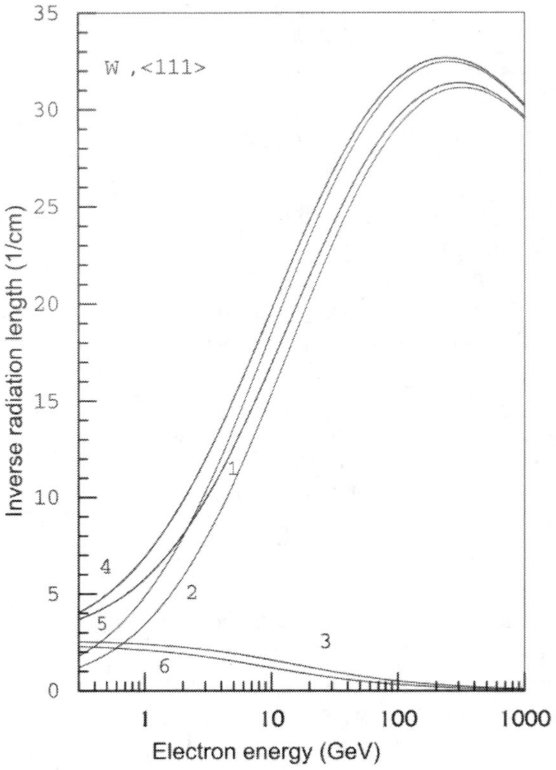

Figure 3. The inverse radiation length in the tungsten crystal, axis $< 111 >$ at different temperatures T vs the electron initial energy. The curves 1 and 4 are the total effect: $L^{cr}(\varepsilon)^{-1} = I(\varepsilon)/\varepsilon$ Eq.(63) for T=293 K and T=100 K correspondingly, the curves 2 and 5 give the coherent contribution $I^F(\varepsilon)/\varepsilon$ Eq.(65), the curves 3 and 6 give the incoherent contribution $I^{inc}(\varepsilon)/\varepsilon$ Eq.(67) at corresponding temperatures T.

The second term of decomposition of Eq.(63)($\propto \nu_0^2$) gives the intensity of incoherent radiation:

$$I^{inc}(\varepsilon) = \frac{\alpha m^2}{60\pi}\frac{\varepsilon}{\varepsilon_0} g \int\limits_0^{x_0} e^{-x/\eta_1} J(\chi)\frac{dx}{x_0}, \tag{67}$$

the new representation of $J(\chi)$ is

$$J(\chi) = \frac{i\pi}{2}\int\limits_{\lambda-i\infty}^{\lambda+i\infty} \frac{\chi^{2s}}{3^s}\frac{\Gamma(1+3s)}{\Gamma(s)} R(s)\frac{ds}{\sin^2\pi s}, \quad -\frac{1}{3} < \lambda < 0, \tag{68}$$

where

$$R(s) = 15 + 43s + 31s^2 + 28s^3 + 12s^4. \tag{69}$$

The inverse radiation length in tungsten crystal (axis $< 111 >$) $1/L^{cr}(\varepsilon) = I(\epsilon)/\varepsilon$ Eq.(63), as well as the coherent contribution $1/L^F(\varepsilon) = I^F(\varepsilon)/\varepsilon$ Eq.(65) and the incoherent contribution $1/L^{inc}(\varepsilon) = I^{inc}(\varepsilon)/\varepsilon$ Eq.(67) are shown in Fig.3 for two temperatures T=100 K and T=293 K as a function of incident electron energy ε. One can see that at temperature T=293 K the intensity $I^{coh}(\varepsilon)$ is equal to $I^{inc}(\varepsilon)$ at $\varepsilon \simeq 0.4$ GeV and temperature T=100 K the intensity $I^{coh}(\varepsilon)$ is equal to $I^{inc}(\varepsilon)$ at $\varepsilon \simeq 0.7$ GeV. At higher energies the intensity $I^{F}(\varepsilon)$ dominates while the intensity $I^{inc}(\varepsilon)$ decreases monotonically.

Interrelation between the theory and experiment (very limited for the time being) is shown in Fig.4. One can see that there is quite satisfactory agrement of the theory and data.

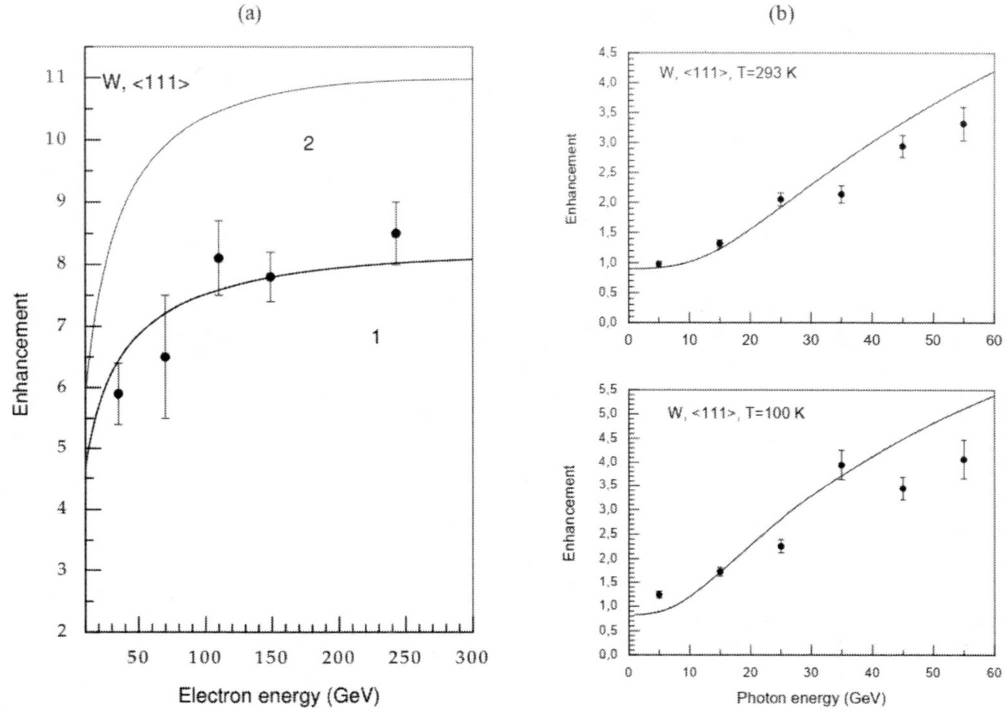

Figure 4. Comparison of theory and experiment.(a) Enhancement of radiation intensity (the ratio L^{BM}/L^{ef}) in the tungsten crystal, axis $< 111 >$, T=293 K. The curve 1 is for the target with thickness $l = 200\ \mu m$, where the energy loss was taken into account. The curve 2 is for a considerably more thinner target, where one can neglect the energy loss ($L^{ef} \to L^{cr}$). The data are from *K.Kirshbom, et al, 2001*(see (BK06)).
(b) Enhancement of the probability of pair creation in the tungsten crystal for different temperatures, axis $< 111 >$. The data are from *K.Kirshbom, et al, 1998* (see (BK05)).

7. Back-reaction: spectra of radiation taking into account energy loss in oriented crystals

The crystal radiation length $L(\varepsilon) = \varepsilon/I(\varepsilon)$, $I(\varepsilon)$ is the intensity of electron radiation, and the pair creation length $L_{pr}(\omega) = 1/W(\omega)$, $W(\omega)$ is the pair creation probability, are the function of energy in oriented crystals.

We consider the case when the target thickness l is of the order $l \sim L(\varepsilon)$ in the intermediate energy region. Here we will neglect the energy dispersion. On this assumption the energy loss equation acquires the form

$$dt = \frac{L(\varepsilon)}{\varepsilon} d\varepsilon, \quad t(\varepsilon, \varepsilon_0) = \int_{\varepsilon}^{\varepsilon_0} \frac{dx}{x} L(x), \quad \varepsilon = \varepsilon(\varepsilon_0, t). \tag{70}$$

Journal of Physics: Conference Series **198** (2009) 012003 doi:10.1088/1742-6596/198/1/012003

Now the photon spectral distribution taking into account the energy loss can be written in the form

$$\omega\frac{dn_\gamma}{d\omega} = \int\limits_0^l \frac{dI(\varepsilon(\varepsilon_0,t),\omega)}{d\omega}\vartheta(\varepsilon(\varepsilon_0,t)-\omega)dt$$
$$= \int\limits_{\varepsilon_l}^{\varepsilon_0} \frac{L(\varepsilon)}{\varepsilon}\frac{dI(\varepsilon,\omega)}{d\omega}\vartheta(\varepsilon-\omega)d\varepsilon, \quad \varepsilon_l=\varepsilon(\varepsilon_0,l), \tag{71}$$

where $dI(\varepsilon,\omega)/d\omega$ is the radiation intensity spectral distribution (see Eq.(63)), $\varepsilon_l = \varepsilon(\varepsilon_0,l)$ is the electron energy after traversing the thickness l by electron with the initial energy ε_0. The result of calculation for the tungsten crystal is shown in Fig.5.

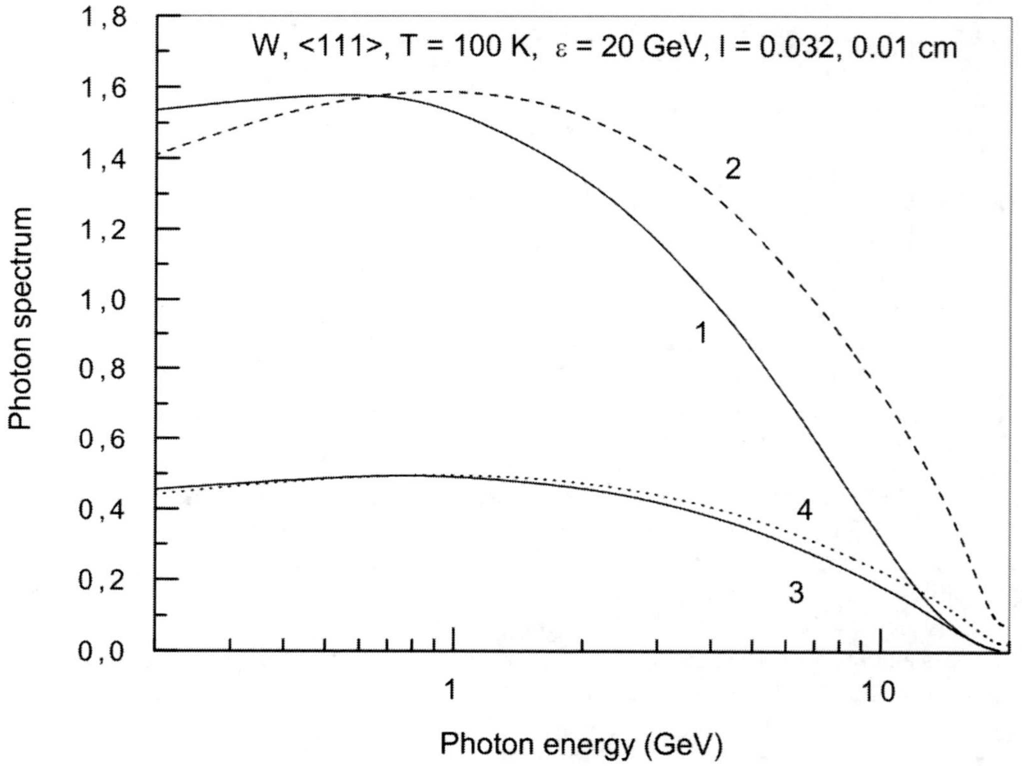

Figure 5. The spectral distribution of radiation at the initial electron energy $\varepsilon_0 = 20$ GeV in the tungsten crystal, axis $< 111 >$, T=100 K in two targets with thickness $l = 0.032$ cm= $0.77\ L = 0.16\ L_{pr}$ (curves 1 and 2) and $l = 0.01$ cm= $0.24\ L = 0.093\ L_{pr}$ (curves 3 and 4) vs the photon energy ω. The curves 2 and 4 are calculated according to Eq.(63), while the curves 1 and 3 are calculated according to Eq.(71) which takes into account the electron energy loss.

Quasiclassical Operator Method: List of references

Books

(1) Baier V N Katkov V M and Fadin V M 1973 (BKF73), *Radiation from Relativistic Electrons* (in Russian) (Moscow: Atomizdat)

(2) Baier V N Katkov V M and Strakhovenko V M 1998 (BKS98), *Electromagnetic Processes at High Energies in Oriented Crystals* (Singapore: World Scientific)

(3) Berestetskii V B Lifshitz E M and Pitaevskii L P 1982 (BLP82) *Quantum Electrodynamics* 2nd ed (Oxford: Pergamon Press)

Articles

(1) Baier V N and Katkov V M 1967 (BKp67) Radiative polarization of electrons in a magnetic field *Soviet Phys. JETP* **25** 944

(2) Baier V N and Katkov V M 1967 (BK67) Quantum effects in a magnetic bremsstrahlung *Phys. Lett.* A **25** 492

(3) Baier V N and Katkov V M 1968 (BK68) Processes involved in the motion of high energy particles in a magnetic field *Soviet Phys. JETP* **26** 854

(4) Baier V N and Katkov V M 1969 (BK69) Quasiclassical theory of bremsstrahlung by relativistic particles *Soviet Phys. JETP* **28** 807

(5) Baier V N 1972 (B72) Radiative polarization of electrons in storage rings *Soviet Phys. Uspekhi* **14** 695

(6) Baier V N, Katkov V M and Strakhovenko V M 1973 (BKS72) Radiation emitted by relativistic particles in periodic structures *Soviet Phys. JETP* **36** 1120

(7) Baier V N and Katkov V M 1976 (BK76) Corrections to the eikonal approximation and the bremsstrahlung of electrons at high energies *Soviet Phys.-Doklady* **21** 150

(8) Baier V N, Katkov V M and Strakhovenko V M 1981 (BKS81) Radiation of relativistic particles moving quasiperiodically *Soviet Phys. JETP* **53** 688

(9) Baier V N, Fadin V S, Khoze V A and Kuraev E A 1981 (BFKK81) Inelastic processes in high energy quantum electrodynamics *Phys. Reports* **78** 293

(10) Baier V N, Katkov V M and Strakhovenko V M 1988 (BKSi88) Incoherent radiation and pair creation in crystals *Phys. Stat. Solidi (b)* **149** 403

(11) Baier V N, Katkov V M and Strakhovenko V M 1988 (BKS88) Radiation from relativistic particles colliding in a medium in the presence of an external field *Soviet Phys. JETP* **67** 70

(12) Baier V N, Katkov V M and Strakhovenko V M 1989 (BKS89) Quantum radiation theory in inhomogeneous external fields *Nucl.Phys.* B **328** 387

(13) Baier V N, Katkov V M and Strakhovenko V M 1989 (BKSp89) Pair production by a photon in inhomogeneous external fields *Phys. Lett.* B **225** 193

(14) Baier V N, Katkov V M and Strakhovenko V M 1990 (BKS90) Quantum beamstrahlung and electroproduction of the pairs in linear colliders *Particle Accelerators* **30** 43

(15) Baier V N, Katkov V M and Strakhovenko V M 1991 (BKS91) Semiclassical theory of electromagnetic processes in a plane wave and a constant field *Soviet Phys. JETP* **73** 945

(16) Katkov V M and Strakhovenko V M 2001 (KS01) Operator representation of the wave function for a charged particle in an external field and its application *JETP* **92** 561

(17) Baier V N and Katkov V M 2004 (BK04) On coherent radiation in electron-positron colliders *Quantum aspects of Beam Physics* ed Chen P and Reil K (Singapore: World Scientific) p 3

(18) Baier V N and Katkov V M 2005 (BKPR05) Concept of formation length in radiation theory *Phys. Reports* **409** 261

(19) Baier V N and Katkov V M 2005 (BK05) Coherent and incoherent pair creation by a photon in oriented single crystal *Phys. Lett.* A **346** 359

(20) Baier V N and Katkov V M 2006 (BK06) Coherent and incoherent radiation from high-energy electron and the LPM effect in oriented single crystal *Phys. Lett.* A **353** 91

(21) Baier V N and Katkov V M 2007 (BK07) Pair creation by a photon in a strong magnetic field *Phys. Rev.* D **75** 073009

(22) Baier V N and Katkov V M 2009 (BK09) Spectra of radiation and created particles at intermediate energy in oriented crystal taking into account energy loss (*Preprint* hep-ph 0901.3951v1)

Joint IPPP, CI and ICFA Workshop on Advanced QED Methods for Future Accelerators IOP Publishing
Journal of Physics: Conference Series **198** (2009) 012004 doi:10.1088/1742-6596/198/1/012004

On the equivalence of semi-classical methods for QED in intense external fields

Anthony Hartin

DESY FLC, Notkestrasse 85, Hamburg, Germany

E-mail: anthony.hartin@desy.de

Abstract. Using the semi-classical method of Nikishov-Ritus (NR), the derivation of the transition rate of the beamsstrahlung process is reviewed. This method uses the Furry Picture and the exact solutions of the Dirac equation in the external field potential. For future linear colliders, the nominal machine parameters are such that this external field can be considered to be a constant crossed electromagnetic field. The Dirac equation solutions can be Fourier transformed such that they are functions of Dirac gamma matrices, Airy functions and the usual non-external field solution. The resultant analytic form for the transition rate is the same as that obtained by the Quasiclassical Operator (QO) method of Baier-Katkov which is valid in the limit of ultra-relativistic electron and vanishingly small radiation angle. The NR calculation however also exhibits a pole in the radiation angle for back-radiated photons. The removal of this pole requires a further study of IR divergences within the Furry Picture.

1. Introduction

There has been a revival in interest over the past decade or so, in fundamental QED processes occurring in intense external fields. A SLAC experiment involving the interaction of an intense laser with an electron beam showed clearly the onset of multiple numbers of laser photons contributing to the process [1]. Around the same time a similar set of experiments in Europe, on electron laser interaction were performed which observed a fermion mass shift due to the action of the electron motion in the laser field [2, 3]. Another set of experiments has taken place at CERN during the last few years involving fermion processes in the intense fields present in the atomic lattices of crystals [5]. Good agreement has been found with theoretical calculations of QED process in crystals [4].

The ratio of the strength of the external field to the Schwinger critical field (at which real e^+e^- pairs can be produced from the vacuum) gives a measure as to whether a quantum or classical treatment of the interaction with the external field is necessary. An 'intense field', then, has a field strength sufficiently high to require a quantum treatment. Several calculation strategies exist.

If the field strength is not too high the external field can be considered to contribute one or more photons in the Born approximation [6]. Higher field strengths require that collective effects be taken into account by performing a QED calculation in the Furry Picture (FP). In this picture the external field potential is bundled up with the Dirac field to form a bound Dirac field which interacts with the free Maxwell field. Exact solutions of the Dirac equation in a (classical) external potential exist and are used to calculate the transition probabilities of interest.

Two calculation strategies are compared in this paper. In the first strategy (also first historically) the transition rates of various first order processes in the FP were calculated using the usual S-matrix theory. The interpretation of the external field as photons contributing to the process emerges naturally. Exact results for transition rates with no kinematic approximations can be obtained. This method is referred to as the Nikishov-Ritus (NR) method in this paper [7, 8].

The second method makes use of the fact that in particle colliders the initial state fermions are ultra-relativistic and the energy levels of the fermion states in the external field of nominally designed bunches are extremely close together. If the fermion final states are also ultra-relativistic then the fermion motion can be considered classical in the external field. This Quasi-classical Operator (QO) method starts from an intermediate point in perturbation theory before the integration with respect to time has been carried out. The transition probability is obtained after allowing the operators of the electron motion to commute [9].

The first order processes in the FP for a number of different external fields, including electromagnetic fields of various polarisations, magnetic fields and intra-crystalline fields, can be considered well studied. Second order processes in the FP require substantial further study, as do the radiative corrections [10]. In this paper the beamsstrahlung process, i.e the bremsstrahlung in the electromagnetic field of a relativistic particle bunch, is considered. The NR method for this process will be reviewed and the transition rate will be compared to that obtained using the QO method.

One application of interest for this work is that of spin tracking through the beam-beam interaction at colliders. The CAIN program [11] which performs this spin tracking, assumes no radiation angle for the outgoing photon. In the interests of precision spin tracking, especially given that collider designs include a crossing angle, it is important to know the variation of calculated transition probability with radiation angle and no kinematic approximations [12]. The full NR result will therefore be examined for this behaviour.

2. QED in intense fields

The motivation for using the FP arose in the studies of electrons in atomic systems in which bound states were natural ones. The interaction of the electrons with the external field can be handled by iteration using the Born approximation. However for strong enough fields the Born series does not converge and the interaction with the external field has to be taken into account to all orders. The FP achieves this by repackaging the Lagrangian density of the system with the interaction term specifying the coupling between free Maxwell and the combined Dirac-external field system. The usual perturbation method within S-matrix theory is then employed using the solution for the bound Dirac field equation,

$$\left[(p - eA^e)^2 - m^2 - \tfrac{ie}{2}F^e_{\mu\nu}\sigma^{\mu\nu}\right]\psi_V(x,p) = 0$$

$$\text{where} \qquad F^e_{\mu\nu} = k_\mu \frac{\partial A^e_\nu}{\partial\phi} - k_\nu \frac{\partial A^e_\mu}{\partial\phi} \qquad \text{and} \qquad \phi = k^\mu x_\mu \tag{1}$$

The solution $\psi_V(x,p)$ to the bound Dirac field equation was worked out long ago [13]. A nice exposition of its derivation is found in section 40 of [14]. This Volkov solution is a product of a Volkov $E_p(x)$ function of the external field 4-potential A^e_μ, and the usual free field solution,

$$\psi_V(x,p) = E_p(x)\ \exp(-ip\cdot x)\ u_s(p)$$

$$E_p(x) = \left[1 + \frac{e}{2(k\cdot p)}\not{k}\not{A}^e\right]\exp\left(-i\int_0^{kx}\left[\frac{e(A^e\cdot p)}{(k\cdot p)} - \frac{e^2A^{e2}}{2(k\cdot p)}\right]d\phi\right) \tag{2}$$

The orthogonality of the Volkov $E_p(x)$ is an important result [15, 16] which allows the bound fermion propagator $G^e(x, x')$ to be written as the usual propagator sandwiched between Volkov $E_p(x)$ functions,

$$G^e(x, x') = \frac{1}{(2\pi)^4} \int_{\infty}^{\infty} d^4p \, E_p(x) \frac{\not{p} + m}{p^2 - m^2} \bar{E}_p(x') \tag{3}$$

Using the above results the required interaction between bound fermions and photons can be studied using the usual S-matrix theory and Wick's theorem. Since the Volkov solution is written in terms of the bispinor and gamma matrices the usual Dirac gamma algebra and spin and polarization sums are used to obtain the transition probability of the desired QED process. This was the route chosen by Nikishov and Ritus (and others) in the 1960s and which is summarised in the following two sections.

With the solutions of the bound Dirac equation and an expression for the bound fermion propagator, Volkov $E_p(x)$ functions can be grouped together to form only one new element in the S-matrix - a modified vertex

$$\Gamma^e_\mu = \int d^4x \; \bar{E}_{p_f}(x) \gamma_\mu E_{p_i}(x) \exp\left[i(p_i - p_f - k_f) \cdot x)\right] \tag{4}$$

In terms of the Feynman diagrams, the bound Dirac equation solutions are represented as double straight lines.

3. Bremsstrahlung in a constant crossed field

The beamsstrahlung process (figure 1) is simply the bremsstrahlung process that takes place when a charge radiates a photon in the field of an oncoming charge bunch in collider collisions. For proposed future linear colliders the radiation coherence length - classically, the ratio of the radius of curvature of the radiating charge to the relativistic gamma value of the radiating charge - is shorter than both the typical bunch length and the characteristic length specified by the beam disruption parameter. Therefore the external field can be considered to be a constant crossed field [17] specified by

$$\begin{aligned} A^e_\mu &= (k \cdot x) \, a_{1\mu} \\ (a_1 \cdot a_1) &= -a^2 \end{aligned} \tag{5}$$

The Lorentz condition and a choice of reference frame such that the zeroth (the time) component of the external field 4-potential is zero, requires that the 3-potential $\vec{a}_1$ and the 3-momentum $\vec{k}$ of the external field be orthogonal. Using these vectors as a basis for a coordinate system with $\vec{a}_1$ forming the x-axis and $\vec{k}$ the z-axis, the radiation angles (θ_f, ϕ_f) of the radiated photon $\vec{k}_f$ can be specified (figure 2). If the radiating electron is directed exactly opposite to the wave-vector of the oncoming bunch field (as is usually the case in colliders except for a small crossing angle) then the angle $\theta_f = \pi$ corresponds to forward radiation and $\theta_f = 0$ to back radiation

An explicit form for the Volkov $E_p(x)$ functions can be obtained by substitution of the 4-potential of the constant crossed field. In order to simplify the dependence on the space-time variable x_μ, a Fourier transform is made to an infinite integration over a variable r

Journal of Physics: Conference Series **198** (2009) 012004 doi:10.1088/1742-6596/198/1/012004

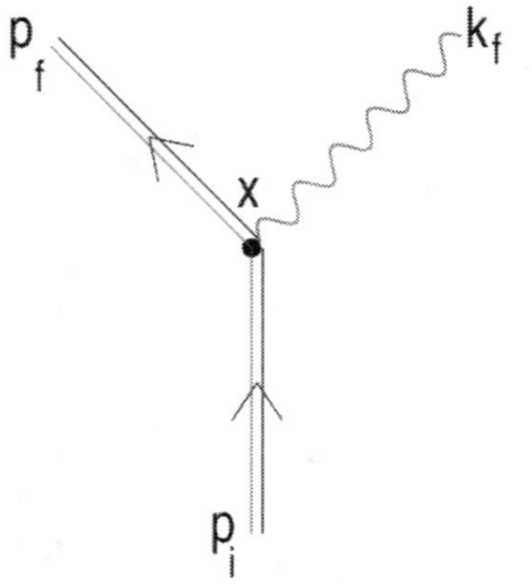

Figure 1. Beamsstrahlung Feynman diagram.

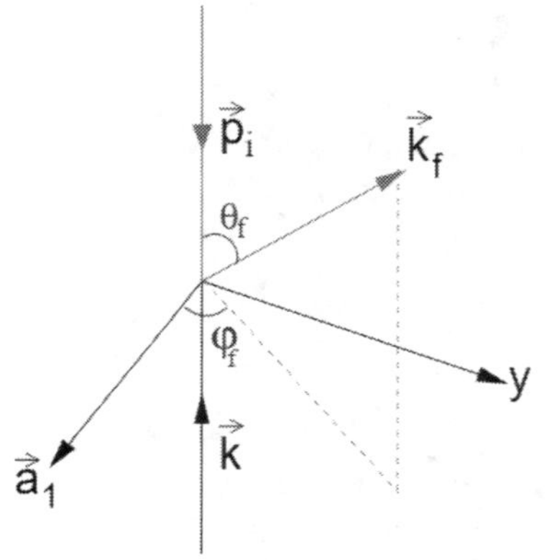

Figure 2. Coordinate system and radiation angles.

$$
\begin{aligned}
(k \cdot x)^n \exp(iS(x)) &= \int_{-\infty}^{\infty} dr\ F_n(r) \exp(-ir(k \cdot x)) \\
\text{where}\quad F_n(r) &= \int_{-\infty}^{\infty} dt\ t^n \exp(irt + iS(t)) \\
\text{and}\quad S(t) &= -\left[\frac{e(a_1 \cdot p)}{2(k \cdot p)} t^2 - \frac{e^2 a^2}{6(k \cdot p)} t^3\right]
\end{aligned} \tag{6}
$$

The modified vertex can likewise be transformed and the integration over space-time carried out to produce the usual delta function whose argument expresses the momentum conservation

$$
\Gamma^e_\mu(r) = (2\pi)^4 \int_{-\infty}^{\infty} dr\ E_p(r)\gamma_\mu E_{p_i}(r)\ \delta^4(p_f + k_f - p_i - rk) \tag{7}
$$

There is a contribution rk_μ from the external field, which is integrated over all values allowed by conservation of momentum. It is open to interpretation whether this is indeed one photon of 4-momentum rk_μ or r photons of 4-momenta k_μ each. This is unlike the case for processes in laser fields in which the contribution from the external field is a summation (over, say, n) rather than an integration and the interpretation clearly is in terms of n external field photons. Though external field photons were not included at the outset, they emerge through the quantum interaction of the electron with the external field.

For a constant crossed field, the concept of a quasi-momentum for the fermions $q_\mu = p_\mu + \frac{e^2a^2}{2(k\cdot p)}\, k_\mu$ is ill defined, since the intensity of the crossed field $\frac{e^2a^2}{m} \to \infty$. Consequently the conservation of momentum of the process is written in terms of the usual external field-free momentum p_μ.

4. Beamsstrahlung transition probability

The matrix element of the beamsstrahlung process is the modified vertex sandwiched between Dirac bispinors. In deriving the differential transition probability, dw, it proved useful to use relativistic normalization for the field operators and to follow the derivation of the general form in [18]. It should be noted that the square of the matrix element will introduce another integration (over variable r') and a second delta function. The particle energy-momenta are specified by $(\epsilon_i, \vec{p_i})$, $(\epsilon_f, \vec{p_f})$ and $(\omega_f, \vec{k_f})$.

Journal of Physics: Conference Series **198** (2009) 012004 doi:10.1088/1742-6596/198/1/012004

$$dw = \frac{1}{16\epsilon_i \pi^2} \overline{\sum_{if}} \left| \overline{u}(p_f)\Gamma^e_\mu(r)u(p_i)\right|^2 \frac{d^3\vec{p}_f}{\epsilon_f}\frac{d^3\vec{k}_f}{\omega_f} \tag{8}$$

There is nothing conceptually difficult in reducing the transition probability to its simplest form, however the procedure is technical and time-consuming. The main steps are simply listed and the end result quoted for final analysis:

- Perform the integration over initial 3-momenta using 3 components of one of the delta functions
- Write the Fourier transformed functions F_n as Airy functions by using the relation

$$\int_{-\infty}^{\infty} dt\ \exp\left(iat + bt^2 + i\frac{c}{3}t^3\right) = c^{-1/3}\mathrm{Ai}\left[c^{-1/3}\left(a - \frac{b^2}{c}\right)\right] \exp\left[i\frac{b}{c}\left(\frac{2b^2}{3c} - a\right)\right]$$

- Perform the spin sums and trace calculation using the usual Dirac algebra
- Introduce an additional delta function using the relation

$$\int \frac{d^3\vec{p}_f}{2\epsilon_f} = \int d^4p_f\ \delta(p_f^2 - m^2)$$

- Using two delta functions do the integrations over final fermion 3-momenta $\vec{p}_f$ and the contribution from the external field r, to get the condition $r \to (p_i \cdot k_f)/(k \cdot p_f)$
- Integrate over the x-component of the final photon momentum k_f^x to get the condition that $r' \to r$ and transform the transition probability w to a transition rate W
- Transform the y-component of the final photon 3-momentum k_f^y and integrate to reduce products of Airy functions to a single Airy function [19] using

$$\int_0^{\infty} \frac{dt}{\sqrt{t}}\mathrm{Ai}^2(t+a) = \frac{1}{2}\int_{2^{2/3}a}^{\infty} \mathrm{Ai}(t)\ dt$$

- Replace Airy functions with Bessel functions using relations like, $\mathrm{Ai}(t) = \frac{\sqrt{t}}{\pi\sqrt{3}}\mathrm{K}_{1/3}\left(\frac{2}{3}t^{3/2}\right)$
- The integration over the z-component of the final photon 3-momenta k_f^z is transformed to a function of scalar products of 4-momenta $u = \frac{(k \cdot k_f)}{(k \cdot p_f)}$

The final result for the transition rate is

$$\frac{dW}{du} = \frac{\alpha m^2}{\pi\sqrt{3}\epsilon_i}\frac{1}{(1+u)^2}\left[\int_\chi^\infty K_{5/3}(y)dy - \frac{u^2}{1+u}K_{2/3}(\chi)\right]$$
$$\text{where} \quad \chi = \frac{2u}{3ea(k \cdot p_i)} \quad \text{and} \quad u = \frac{(k \cdot k_f)}{(k \cdot p_i) - (k \cdot k_f)} \tag{9}$$

Journal of Physics: Conference Series **198** (2009) 012004 doi:10.1088/1742-6596/198/1/012004

5. Comparison of the Nikishov-Ritus and Baier-Katkov methods

The same beamsstrahlung transition rate appears in the literature for both NR and QO methods, though the exact form of the integration variable varies from derivation to derivation and NR prefers to write in terms of Airy functions. Nevertheless, a direct comparison can be made and the difference in the final result for the two methods is encapsulated in the integration variable χ,

$$\chi(\mathrm{QO}) = \frac{1}{3ea(k\cdot p_i)}\frac{\omega_f}{\epsilon_i-\omega_f}$$
$$\chi(\mathrm{NR}) = \frac{2}{3ea(k\cdot p_i)}\frac{(k\cdot k_f)}{(k\cdot p_i)-(k\cdot k_f)} \tag{10}$$

Physically, the parameter χ is inversely proportional to the magnetic field strength of the oncoming charge bunch and proportional to the ratio of radiated photon energy to final electron energy. The $\chi(\mathrm{NR})$ also retains an angle dependence in its scalar products. Naively, if the electron is ultra-relativistic and radiates almost entirely in the forward direction $\theta_f = \pi$ (as one might expect) then $\chi(\mathrm{NR}) \to \chi(\mathrm{QO})$ and the two results seem consistent in the limit assumed by the QO method.

However, the beamsstrahlung transition rate contains a pole at the lower end of the χ integration (at $\chi = 0$) since the K Bessel function of fractional order approaches infinity as its argument approaches zero [20]. Comparison between the QO and NR methods near the pole requires careful consideration. Both $\chi(\mathrm{NR})$ and $\chi(\mathrm{QO})$ tend towards zero as the energy of the radiated photon becomes small. This is the IR divergence. Applying the usual argument for the removal of the divergence, a cut is made at a minimum energy which is the limit of the energy resolution of the detector. The residual divergent pole is then removed theoretically by inclusion of radiative corrections, though it should be noted that it does not appear to have been done explicitly in the literature for this process in the FP.

Writing the minimum photon energy cut instead as a fictious small photon mass m_γ the parameter $\chi(\mathrm{NR})$ can be written in terms of the radiation angle

$$\chi(\mathrm{NR}) = \frac{2}{3ea(k\cdot p_i)}\frac{\sqrt{|\vec{k}_f|^2+m_\gamma^2}-|\vec{k}_f|\cos(\theta_f)}{2\epsilon_i-\sqrt{|\vec{k}_f|^2+m_\gamma^2}-|\vec{k}_f|\cos(\theta_f)} \tag{11}$$

and its dependence plotted (figure 3(a)). The closer $\chi(\mathrm{NR})$ reaches zero, the more closely the transition rate pole is reached. Clearly this is the case for $\theta_f = 0$, i.e. back radiation and is the case even for ultra-relativistic electrons. This is not a trivial observation since even with an energy cut, the transition rate reaches a maximum around the pole and any small variation at that point will have a relatively large effect on the transition rate value.

A maximum transition rate for back radiation is obviously in contradiction to the results of classical electrodynamics which predict forward radiation for ultra relativistic electrons. Additionally the pole in the transition rate for back radiation renders the equivalence of the QO and NR results within the ultra relativistic limit invalid.

In principle, the removal of the IR divergence from the beamsstrahlung process within the FP should be straightforward. It has been argued that since IR divergences cancel in the normal interaction picture, then it must also be the case for the bound interaction picture [22]. The removal may require a calculation of the vertex function in the FP. In any case a dedicated further study is necessary.

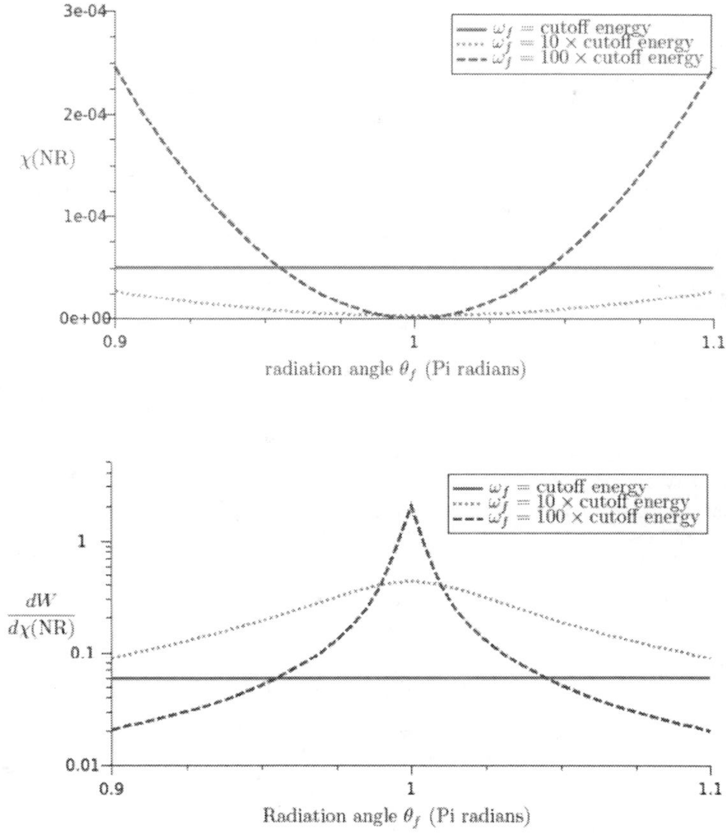

Figure 3. Differential transition rate $\frac{dW}{d\chi(\text{NR})}$ and parameter $\chi(\text{NR})$ near the transition rate pole.

6. Conclusion

The analytic form of the transition rate of the beamsstrahlung process has been compared for two calculation methods. The QO method assumes a classical interaction between the electron and the external field, but a quantum interaction to produce the radiated photon. The spectrum of the radiation (the derivative of the transition rate with respect to the radiated photon energy) shows the usual IR divergence for low energy photons. The QO calculation is only expected to be valid for an ultra-relativistic electron before and after the radiation.

The NR method is a full calculation using the bound interaction picture within the usual S-matrix theory with no kinematic approximations. In certain limits the analytic form of the NR radiation intensity appears to reduce to that of the classical Schott formula [21] and to the beamsstrahlung transition rate obtained by the QO method. However a full consideration of the NR calculation reveals a transition rate maximum for back-radiated photons. It seems clear that an explicit inclusion of radiative corrections within the FP is required and such a work is intended for a forthcoming paper.

Acknowledgments
This work has been supported by the DFG via the Emmy-Noether grant Li 1560/1-1.

Journal of Physics: Conference Series **198** (2009) 012004 doi:10.1088/1742-6596/198/1/012004

References

[1] C. Bamber *et al.*, Phys. Rev. D **60**(9) (1999) 092004
[2] D.D. Meyerhofer *et al.*, Phys. Rev. Lett **74** (1995) 2439
[3] D.D. Meyerhofer *et al.*, J. Opt. Soc. Am. B **13** (1996) 113
[4] V.N. Baier, V.M. Katkov and V.M. Strakhovenko, Phys. Lett. **95A** (1983) 403-406
[5] U.I. Uggerhoj, Rev. Mod. Phys. **77** (2005) 1131
[6] D. Schröder, Ph.D Thesis (SLAC Report 371)
[7] A.I. Nikishov and V.I. Ritus, Sov. Phys. JETP **19** (1964) 529-541
[8] A.I. Nikishov and V.I. Ritus, Sov. Phys. JETP **19** (1964) 1191-1199
[9] V.N. Baier and V.M. Katkov, Sov. Phys. JETP **25** (1967) 944-947
[10] A. Hartin, Ph.D Thesis, University of London, 2006
[11] K. Yokoya, CAIN Manual, Version 2.35 (2003) 944-947
[12] I.R. Bailey, A.F. Hartin *et al.*, EUROtev-Report-2008-026
[13] D.M. Volkov, Z. Phys. **94** (1935) 250-260
[14] V.B. Berestetskii, E.M. Lifshitz and L.P. Pitaevskii, *Quantum Electrodynamics*, Pergamon Press 1982
[15] D-S. Guo, T. Åberg, J. Phys. B **24** (1991) 349-359
[16] S. Zakowicz, J. Math Phys. B **46** (2005) 032304
[17] K. Yokoya and P. Chen, Lecture Notes in Physics 400 *Frontiers of Particle Beams: Intensity Limitations*, pp415-445, Springer Verlag 1982
[18] M.E. Peskin and D.V. Schroder, *An introduction to Quantum Field Theory*, Perseus Books 1995
[19] D.E. Aspnes, Phys. Rev. **147**(2) (1966) 554
[20] M. Abromovitz and I. Stegun, *Handbook of Mathematical Functions*, Dover Publications 1964
[21] A.I. Nikishov and V.I. Ritus, Sov. Phys. JETP **25**(6) (1967) 1135
[22] F. Rohrlich, Phys. Rev. **98**(1) (1955) 181

Journal of Physics: Conference Series **198** (2009) 012005 doi:10.1088/1742-6596/198/1/012005

Compton scattering at high intensities

Thomas Heinzl

University of Plymouth, School of Mathematics and Statistics, Drake Circus, Plymouth PL4 8AA, UK

E-mail: thomas.heinzl@plymouth.ac.uk

Abstract. High-intensity Compton scattering takes place when an electron beam is brought into collision with a high power laser. We briefly review the main intensity signatures using the formalism of strong-field quantum electrodynamics.

1. Introduction

Strong field quantum electrodynamics (QED) arises as a slight modification of ordinary QED upon adding a strong *external* electromagnetic field. Traditionally, this would have been the static Coulomb field of a nucleus or a heavy ion leading to effects such as Bremsstrahlung, pair creation [1] or Delbrück scattering [2]. In recent years, huge, time-dependent electromagnetic fields have been produced using high-power lasers, and this will be the focus of this presentation. The magnitudes of some characteristic physical quantities assumed for the purposes of this talk are given in Table 1.

Table 1. Some typical magnitudes characterising current high-power lasers.

Quantity	Magnitude
Power	$P \gtrsim 10^{15}\,\mathrm{W} = 1\,\mathrm{PW}$
Intensity	$I \gtrsim 10^{22}\,\mathrm{W/cm^2}$
Electric field	$E \gtrsim 10^{14}\,\mathrm{V/m}$
Magnetic field	$B \gtrsim 10^{10}\,\mathrm{G} = 10^6\,\mathrm{T}$

Within the next few years one expects to achieve intensities of 10^{23} W/cm^2 at the Vulcan 10 PW upgrade facility [3] and possibly 10^{26} W/cm^2 once the European Extreme Light Infrastructure (ELI) is realised [4]. The currently achieved field strengths should be compared with the genuine QED electric field first obtained by Sauter [5],

$$E_S = \frac{m^2}{e} = 1.3 \times 10^{18}\,\mathrm{V/m}\,, \tag{1}$$

where m is the electron mass and e its charge. The field strength (1) implies an intensity of about 10^{30} W/cm^2. For such extreme magnitudes it becomes energetically favourable for the

Journal of Physics: Conference Series **198** (2009) 012005 doi:10.1088/1742-6596/198/1/012005

vacuum to 'break down' by producing electron positron pairs in order to screen and hence lower the external field (see e.g. [6] for a lucid discussion). The process involved may be interpreted as a nonperturbative tunneling phenomenon and hence is exponentially suppressed [7, 8, 9]. Although there is a huge kinematic pre-exponential factor which somewhat compensates the exponential suppression, current estimates do not predict these 'Schwinger' pairs below 10^{27} W/cm^2. Technically, pair production is signalled by the appearance of an imaginary part in the expression for vacuum polarisation (dressed by the strong external field). The real part, however, is interesting as well as it describes the propagation of photons modified by vacuum polarisation effects, which become magnified and directional due to the external fields present. Possibly most striking is the effect of 'vacuum birefringence', i.e. the occurrence of a phase retardation due to two different and nontrivial refractive indices for different polarisation directions. This was first discussed in Toll's unpublished thesis [10] (see also [11, 12]) and was recently reinvestigated in the light of new laser developments [13] (see also [14]). To enhance the birefringence signal, the strong fields should be probed by an X-ray beam. With current X-ray polarimetry techniques one requires an intensity of at least 10^{25} W/cm^2 to achieve measurable magnitude for the phase retardation effect.

The upshot of this discussion is that strong field vacuum polarisation effects remain elusively small until one gets sufficiently close to the critical field strength (1). Alternatively, one may leave the regime of low energy and use probes (photons or electrons) with energies comparable to the electron mass m or larger. There is, however, a strong field process where intensity effects may readily be studied at low energies and current intensities as given in Table 1. This is high-intensity (or nonlinear) Compton scattering to which we now turn our attention, reviewing our recent analysis [15].

2. Nonlinear Compton scattering: Overview

Our goal is to identify and isolate intensity effects on the spectrum of photons emitted in collisions of a high-power laser with an electron beam. To this end it is useful to define a dimensionless measure of laser intensity via

$$a_0 \equiv \frac{eE\lambdabar}{m} = \frac{eE}{\omega m} , \tag{2}$$

where E is the r.m.s. electric field of the laser having wavelength λ ($\lambdabar \equiv \lambda/2\pi = 1/\omega$). Note that a_0 is a purely classical ratio measuring the energy gain of an electron travelling a distance $\lambdabar$ in the laser field in units of its rest energy, m. Thus, intensity is determined by means of a *probe*, chosen to be an electron. Once a_0 exceeds unity the quiver motion of the probe electron within the beam becomes relativistic. From a particle physics perspective the definition (2) appears gauge invariant as it only involves the field strength, but Lorentz invariance seems lost. The latter may be recovered by writing $a_0^2 = -e^2 \mathsf{A}^2/m^2$ [16, 17] with $\mathsf{A}^2 = \mathsf{A}_\mu \mathsf{A}^\mu$ being the squared gauge potential (suitably time averaged), but this obviously sacrifices gauge invariance. The way out is provided by the Lorentz and gauge invariant definition [18]

$$a_0^2 \equiv \frac{e^2}{m^2} \frac{\langle p_\mu T^{\mu\nu} p_\nu \rangle}{(k \cdot p)^2} , \tag{3}$$

in terms of the electron and photon four-momenta, p and k, and the Maxwell energy momentum tensor, $T^{\mu\nu}$. The brackets $\langle\, . \,\rangle$ denote a Lorentz invariant average over proper time. Note that in the electron rest frame where $p = (m, \mathbf{0})$ we indeed reproduce (2) using $T^{00} = E^2$. We conclude these preliminary remarks, by providing an overview of current and planned laser facilities and their associated intensities in Table 2.

Table 2. Absolute and normalised intensities of current and planned laser facilities. (XFEL: X-ray free electron laser at DESY, Hamburg, FZD: Forschungszentrum Dresden-Rossendorf, Vulcan at Central Laser Facility, Rutherford laboratory [3], ELI: Extreme Light Infrastructure project [4], HiPER: High Power laser Energy Research facility project [19])

	XFEL ('goal')	FZD Draco	Vulcan (1 PW)	Vulcan (10 PW)	ELI/HiPER
I [W/cm^2]	10^{27}	10^{21}	10^{22}	10^{23}	10^{26}
a_0	10	20	70	200	5×10^3

Note that $a_0 \gg 1$ is characteristic for *optical* lasers as the larger frequency of an XFEL reduces the a_0 value according to (2). Thus, henceforth we will assume a scenario where an electron beam collides with photons from an optical laser. To further simplify our analysis we consider a head-on collision only, as schematically sketched in Fig. 1. Before reviewing the strong-field

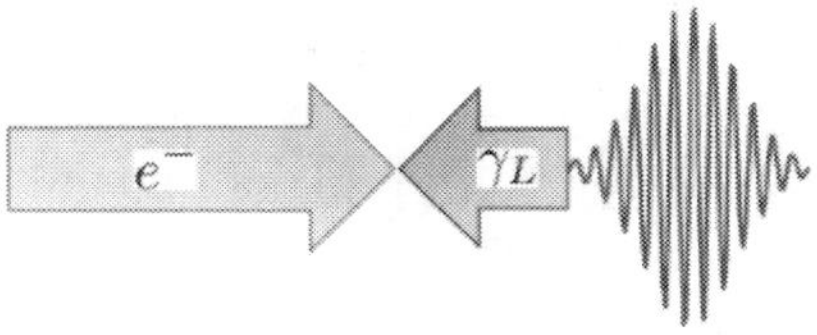

Figure 1. Schematic scenario for nonlinear Compton scattering: an electron beam (e^-) is brought into collision with a laser (γ_L). We concentrate on the special case of a head-on collision.

QED analysis of such a process let us briefly discuss the main intensity effects classically. In principle one can solve for the classical trajectory of the electron subjected to the beam (taken to be a circularly polarised plane wave) and plug the result into the classical formula for the intensity distribution of the ensuing radiation [20]. As the orbits in question are circular one recovers the spectrum first obtained by Schott almost a century ago [21] (see also [22]). The basic intensity effect may already be seen from the first integral of the equation of motion, $p^\mu(\tau)$. Taking the proper time average one finds a mean or quasi momentum given by

$$q^\mu \equiv \langle p^\mu(\tau) \rangle = p^\mu + \frac{a_0^2 m^2}{2\, k \cdot p}\, k^\mu \equiv p^\mu + q_L^\mu \, , \tag{4}$$

which explicitly depends on the intensity parameter a_0 and squares to an effective electron mass

$$q^2 \equiv m_*^2 = m^2(1 + a_0^2) \, , \tag{5}$$

as first obtained by Sengupta [23, 24] and thoroughly discussed in [25]. The zero component of (4), i.e. the average energy shift, was found previously by Volkov upon solving the Dirac equation in a plane wave [26] described by a gauge potential $\mathsf{A}_\mu = \mathsf{A}_\mu(\xi)$, $\xi = k \cdot x$, with k being the four-momentum of the wave. The Volkov solution is then given by the four-spinor,

$$\Psi_{\boldsymbol{p}}(x) = \exp\left\{ - ip \cdot x + \frac{1}{2ik \cdot p} \int^{k \cdot x} d\xi \Big(2ep \cdot \mathsf{A}(\xi) - e^2 \mathsf{A}^2(\xi) \Big) \right\} \chi_{\boldsymbol{p}} \, , \tag{6}$$

where $\chi_{\boldsymbol{p}}$ is a combination of Dirac matrices acting on the standard free Dirac spinor u, see e.g. [15, 17]. The exponent is the classical Hamilton-Jacobi action for an electron in a plane wave. Assuming circular polarisation, $\mathsf{A}^\mu = a_1^\mu \cos(k \cdot x) + a_2^\mu \sin(k \cdot x)$, with $k^2 = 0 = a_i \cdot k$ and $a_i \cdot a_j = -a^2 \delta_{ij}$, the Volkov solution (6) becomes

$$\Psi_{\boldsymbol{p}}(x) = \exp\left\{ -iq \cdot x - ie\frac{a_1 \cdot p}{k \cdot p} \sin(k \cdot x) + ie\frac{a_2 \cdot p}{k \cdot p} \cos(k \cdot x)\right\} \chi_{\boldsymbol{p}} \, . \tag{7}$$

We see the quasi-momentum q making its appearance in the exponent, its longitudinal part q_L stemming from the zero momentum mode of A^2 in (6), $q_L \sim \langle \mathsf{A}^2 \rangle \sim a_0^2$.

The Volkov solution (6) is a crucial ingredient for the formalism of strong-field QED in plane wave backgrounds A based on what is called the Furry picture [27]. Basically, this amounts to altering the decomposition of the Hamiltonian into free and interacting parts by including the background field into the former. In this way, the background is treated exactly while the fluctuations around it are taken into account perturbatively. In this vein, one replaces standard ('bare') fermion lines in Feynman diagrams by *dressed* ones. For external lines, free Dirac wave functions like e.g. $ue^{ip \cdot x}$ are replaced by Volkov wave functions, (6) or (7). Pictorially, this is shown in Fig. 2.

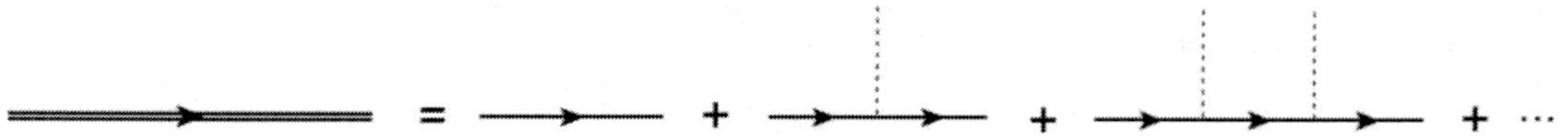

Figure 2. Dressed fermion line expanded in terms of bare ones. The external plane wave field is given by the dotted lines representing continuous absorption/emission of laser photons.

Let us now specialise to nonlinear Compton scattering where a photon γ is emitted upon collisions of electron and laser beams. The Furry picture diagram may be expanded into a series of subprocesses where n laser photons γ_L are being absorbed,

$$e + n\gamma_L = e' + \gamma \, , \tag{8}$$

as shown in Fig. 3.

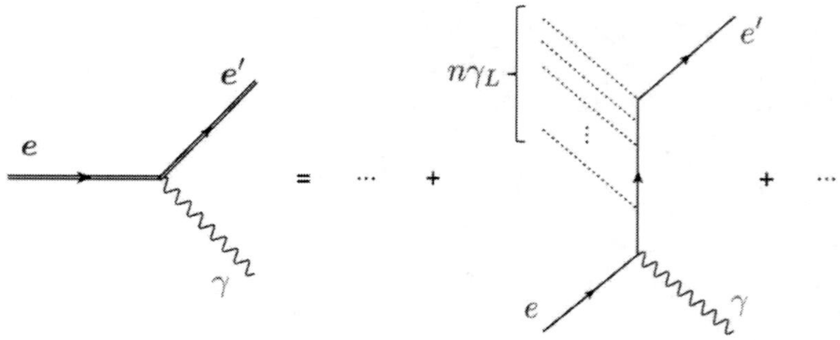

Figure 3. Feynman diagrams for nonlinear Compton scattering. The Furry picture diagram on the left (with dressed electrons) is expanded into an infinite series of subprocesses, $e + n\gamma_L = e' + \gamma$.

The left-hand side of Fig. 3 represents the Furry picture S-matrix element [28, 29, 30]

$$S_{fi} \sim -ie \int \bar{\Psi}_{\boldsymbol{p}'} \not{A} \Psi_{\boldsymbol{p}} \, , \tag{9}$$

which indeed is built from external Volkov lines (so is exact in the background), but otherwise is first order in the fluctuation field A representing the emitted photon γ. Evaluating this expression one recovers the diagrams on the right-hand side of Fig. 3, however, with a nonstandard momentum assignment. The electrons carry quasi-momenta q and q' rather than the asymptotic momenta p and p' in the absence of the field. Accordingly, momentum conservation for each sub-process reads

$$q + nk = q' + k' , \tag{10}$$

with k and k' denoting the momenta of laser and emitted photons, respectively. The processes (8) have been observed before in the experiment SLAC E-144 [31] where a Terawatt laser beam was brought into collision with the 50 GeV electron beam provided by SLAC. As a result the emitted photons had a rather high energy of about 30 GeV. However, only the energy spectra of the outgoing electrons have been recorded and were found to be in agreement with the theory employing the Furry picture (9) as summarised in [16]. The emission rates and spectral distributions of the outgoing photons have not been measured yet but there are plans to do so in the near future at Daresbury laboratories and FZD. Both facilities are working at comparatively low energies, with electron beams in the few 10 MeV range and intensities implying a_0 values of order 10. It is important to note that this regime of strong-field QED (low energies, large intensities) has not been tested before. The expected outcome of these experiments is the topic of the next section.

3. Nonlinear Compton scattering: Results

Before discussing photon emission rates and spectra a few words on the kinematics are in order. Let us first recall the Compton formula for the scattered frequency in the lab frame,

$$\omega' = \frac{\omega}{1 + (\omega/m - \sinh\zeta)\, e^{-\zeta}\, (1 - \cos\theta)} , \tag{11}$$

where the electrons are characterised by rapidity ζ via $e^\zeta = \gamma(1+\beta)$ (β and γ being the usual Lorentz factors) and θ is the usual scattering angle (between $\boldsymbol{k}$ and $\boldsymbol{k}'$). When we generalise this to the nonlinear processes (8) the intensity dependence of the quasi-momenta in (10) is inherited by the scattered frequency such that (11) is superseded by

$$\omega' = \frac{n\,\omega}{1 + \kappa_n\, e^{-\zeta}\, (1 - \cos\theta)} . \tag{12}$$

Here we have defined the dimensionless projection of the total momentum $\boldsymbol{P} \equiv \boldsymbol{q} + n\boldsymbol{k}$ onto the direction $\hat{\boldsymbol{k}}$,

$$\kappa_n = -\boldsymbol{P}\cdot\hat{\boldsymbol{k}}/m = n\omega/m - \sinh\zeta + a_0^2 e^{-\zeta}/2 . \tag{13}$$

Comparing with ordinary Compton scattering (11) we see that (i) the laser frequency ω is replaced by $n\omega$ and (ii) there is a new, intensity dependent addition signalled by the presence of a_0. Item (i) is often referred to as 'higher harmonics generation' and has indeed been observed before by studying the azimuthal intensity distribution of radiation using *linearly* polarised laser light. Second and third harmonics could be identified via their typical quadrupole and sextupole patterns [32, 33].

The main effect, however, to be inferred from (12) is a shift of the Compton edge. To see what is involved let us first consider linear Compton scattering (11). For sufficiently large electron energies ($p^0 \gg m$ or $\gamma \gg 1$) the maximum frequency is obtained for backscattering ($\theta = \pi$), namely

$$\omega'_{\rm max} \simeq 4\gamma^2\omega , \tag{14}$$

which corresponds to a *blue-shift* compared to the incoming laser frequency, $\omega' > \omega$. This is often referred to as 'inverse Compton scattering': in the lab frame the electrons transfer energy to the photons[1]. In the nonlinear case, for the same kinematic situation and $a_0 \gg 1$, the edge value (14) is replaced by the intensity dependent answer,

$$\omega'_{n,\max} \simeq 4\gamma^2 n\omega / a_0^2 \,. \tag{15}$$

Thus, while there is still an overall blue-shift (as long as $a_0 \lesssim 2\gamma$) the result (15) implies an intensity dependent red-shift of the $n = 1$ Compton edge (14) according to

$$\omega'_{\max} \simeq 4\gamma^2 \omega \to 4\gamma^2 \omega / a_0^2 \,. \tag{16}$$

Intuitively this may be explained as follows. Because of its larger effective mass, $m_* > m$, the electron is more static and recoils less so that the energy transfer to the photons is reduced. Thus, already at this stage we can see that increasing intensity produces a tendency to go from inverse to ordinary Compton scattering. This frame dependent effect will be quantified in more detail below. Before that, however, let us have a Lorentz invariant look at the photon emission spectra. To this end we define the invariant parameter

$$x \equiv \frac{k \cdot k'}{k \cdot p'} = \frac{n\nu - \nu'}{\kappa_0 + \nu'} \,, \tag{17}$$

where the second identity holds for a head-on collision in the lab frame with $\nu \equiv \omega/m$ and $\nu' \equiv \omega'/m$. Hence, for the parameter regime we are interested in x is basically proportional to scattered frequency [15]. Fig. 4 shows the resulting spectra as function of x for different intensities a_0. One can clearly see the red-shift (16) of the Compton edge compared to the linear Compton edge ($n = 1$, $a_0 = 0$). With increasing a_0 the edge moves further and further to the left to smaller and smaller x i.e. frequency ω'. In addition, there are side maxima to the right of the $n = 1$ edges, signalling the appearance of higher harmonics with $n > 1$.

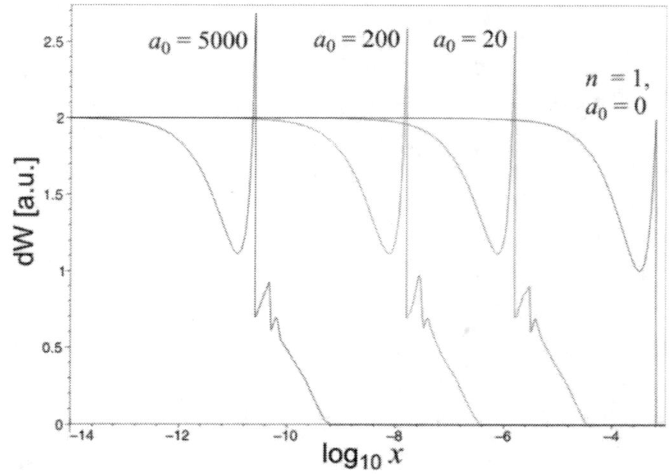

Figure 4. Nonlinear Compton spectra as functions of the invariant x for different values of a_0 and electron energy 40 MeV ($\gamma = 80$, $\nu = \omega/m = 2 \times 10^{-6}$).

If we move on to discussing spectra as a function of scattered frequency ν' there appears a rather rich structure as we tune intensity a_0, displayed in Fig. 5.

[1] Recall that Compton in his original experiment had observed a red-shift upon scattering X-rays off electrons at rest [34].

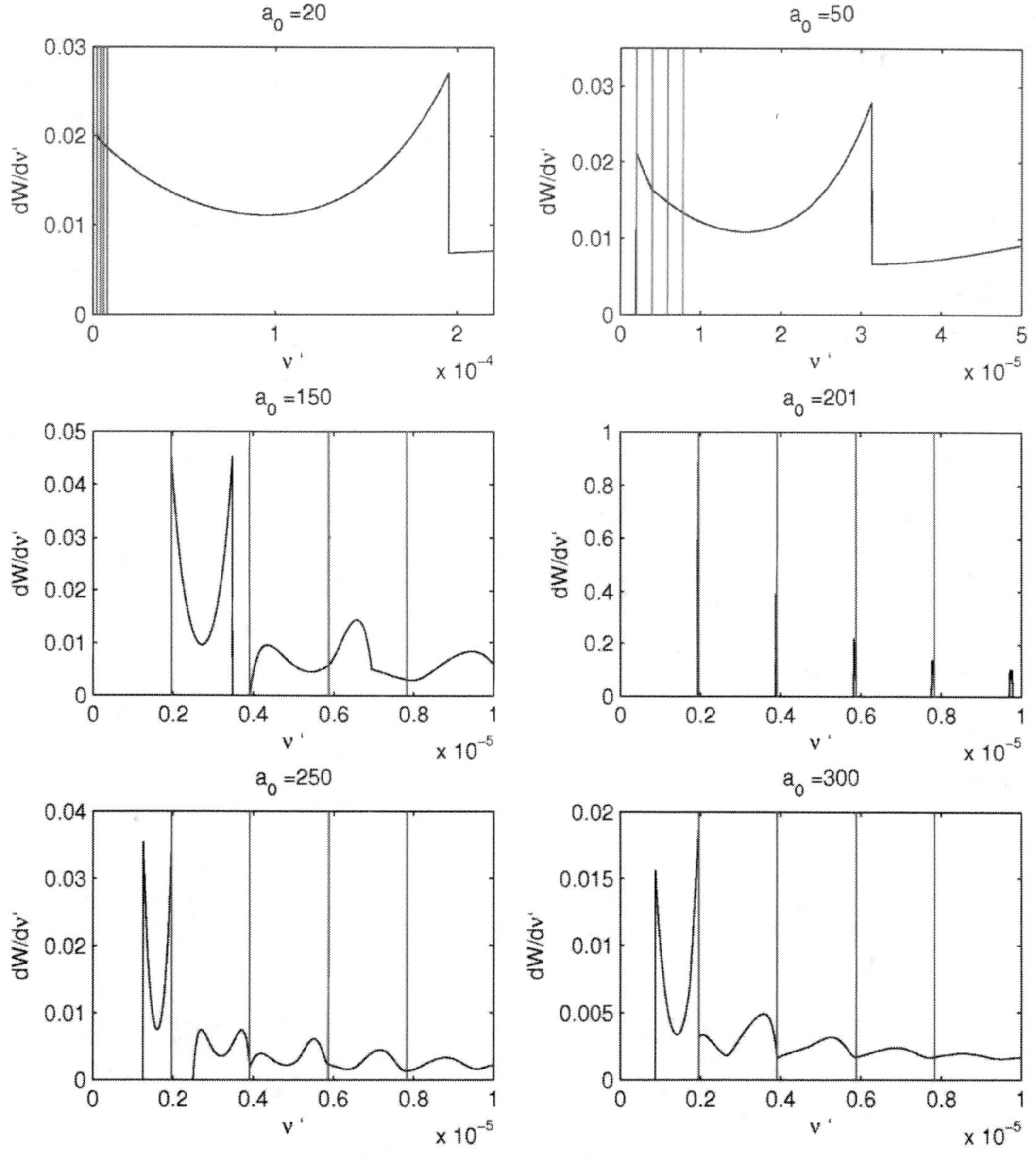

Figure 5. Dependence of the energy spectrum on a_0. The equidistant vertical (red) lines correspond to harmonic frequencies $\nu_n' = n\nu$ (parameters: $\gamma = 100$, $\nu = 2 \times 10^{-6}$).

The spectrum changes significantly upon increasing a_0. For low a_0 values the individual harmonic spectra are all blue-shifted, $\nu_n' > n\nu$, i.e. located to the right of the vertical (red) lines. At $a_0 = 150$ we see a gap appearing between the first and second harmonic and at a critical value, $a_0 \simeq 2\gamma$, the spectrum collapses to (almost) a line spectrum peaked at $\nu_n' = n\nu$, at least for the low harmonics displayed (fourth panel of Fig. 5). Increasing a_0 further the previous blue-shifts turn into red-shifts, $\nu_n' < n\nu$. In other words, inverse Compton scattering changes into Compton scattering with the boundary being given by the appearance of a line spectrum with fixed values for the scattered frequency, $\nu_n' = n\nu$. One may say that this defines an intensity dependent centre-of-mass system where $a_0 \simeq 2\gamma$. This is further elucidated in Fig. 6 which shows the relation between the relevant three-momenta associated with the spectra of Fig. 5.

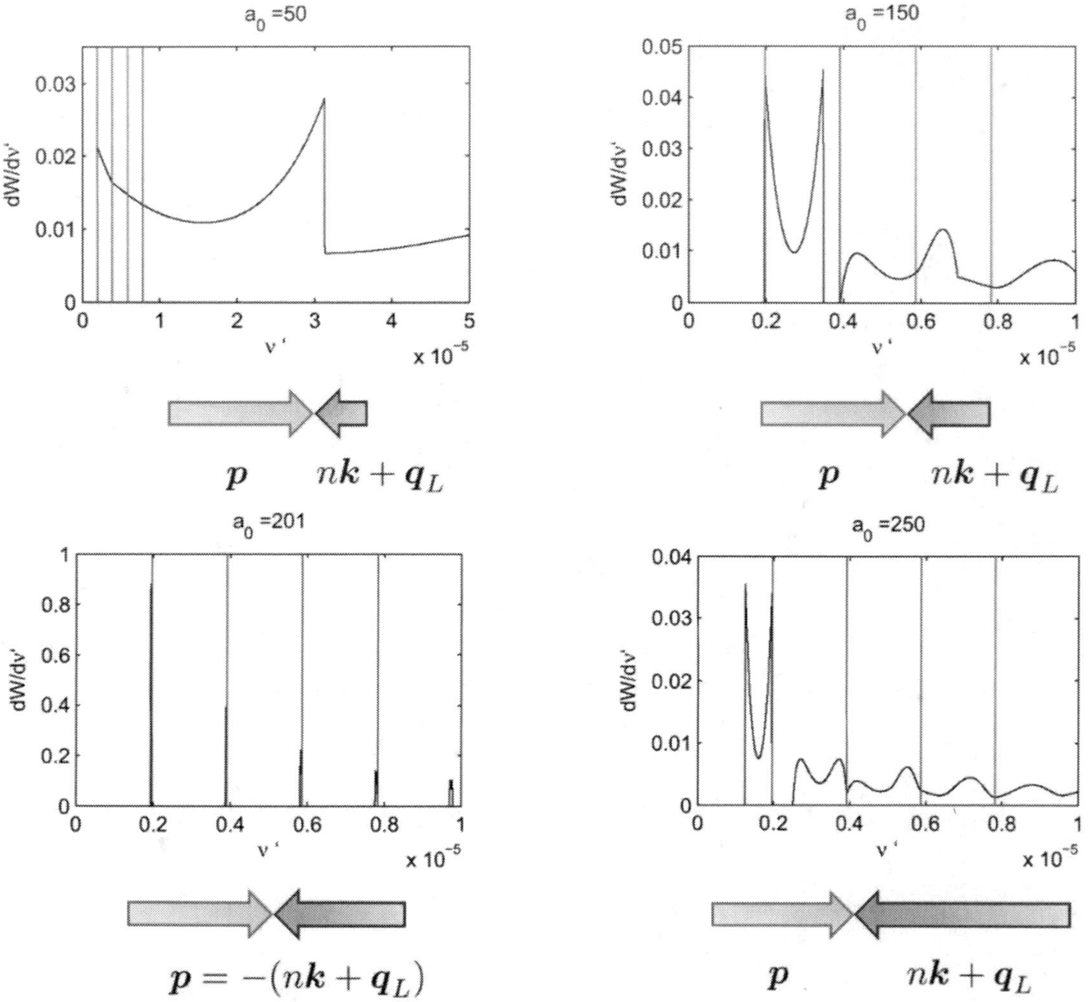

Figure 6. Illustration of the intensity dependent centre-of-mass frame where the total momentum changes direction (third panel). For inverse Compton scattering the electron momentum $\boldsymbol{p}$ dominates in magnitude implying a blue-shift of the scattered frequencies (upper panels). A red-shift results when $|n\boldsymbol{k} + \boldsymbol{q}_L| > |\boldsymbol{p}|$ (last panel).

The magnitude of the asymptotic electron momentum is fixed by the beam specifications, $|\boldsymbol{p}| = me^{\zeta}$. The laser contributes a momentum $n\boldsymbol{k} + \boldsymbol{q}_L$ along $\boldsymbol{k}$ to the total momentum, where $n\boldsymbol{k}$ is the laser momentum proper and the addition $\boldsymbol{q}_L$ changes $\boldsymbol{p}$ into the quasi-momentum $\boldsymbol{q}$, recall (4). This contribution grows linearly in magnitude with a_0, increasingly counterbalancing the incoming electron momentum $\boldsymbol{p}$ until exact compensation is achieved at $a_0 \simeq 2\gamma$. At this point the total momentum $\boldsymbol{P}$ changes direction, and the blue-shift turns into a red-shift.

4. Conclusion

We have given a short overview of the intensity signatures to be expected in the spectra of photons emitted in nonlinear Compton scattering. The main effects are: (i) a red-shift of the linear Compton edge implying a reduction in scattered frequency by a factor of approximately $1/a_0^2$, and (ii) the appearance of smaller side peaks corresponding to higher harmonics. The energy spectra evaluated in the lab frame depend quite sensitively on the value of a_0. Tuning a_0 from small to large values the individual harmonic spectra turn from being blue-shifted (inverse Compton scattering) to red-shifted. At the transition point, $a_0 \simeq 2\gamma$, the spectrum collapses to a line spectrum with peaks precisely at $\nu'_n = n\nu$.

Future work should address the use of more realistic beam shapes (in particular pulsed ones), processes related via crossing symmetry (pair creation and annihilation) and the influence of radiative corrections.

Acknowledgments

The author thanks Chris Harvey and Anton Ilderton for a fruitful collaboration and the organisers of the Workshop on Advanced QED Methods for Future Accelerators for their efforts resulting in such a stimulating environment.

References

[1] Bethe H and Heitler W 1934 *Proc. Roy. Soc. Lond.* **A146** 83–112
[2] Delbrück M 1933 *Z. Phys.* **84** 144
[3] The Vulcan 10 Petawatt Project: http://www.clf.rl.ac.uk/Facilities/vulcan/projects/10pw/10pw_index.htm
[4] The Extreme Light Infrastructure (ELI) project: http://www.extreme-light-infrastructure.eu
[5] Sauter F 1931 *Z. Phys.* **69** 742–764
[6] Coleman S R 1976 *Ann. Phys.* **101** 239
[7] Schwinger J S 1951 *Phys. Rev.* **82** 664–679
[8] Brezin E and Itzykson C 1970 *Phys. Rev.* **D2** 1191–1199
[9] Popov V S 1971 *Pisma Zh. Eksp. Teor. Fiz.* **13** 261–263 [JETP Lett. **13** 185 (1971)]
[10] Toll J 1952 PhD thesis, Princeton, 1952
[11] Narozhnyi N B 1968 *Zh. Eksp. Teor. Fiz.* **55** 714–721 [Sov. Phys. JETP **28**, 371 (1969)]
[12] Brezin E and Itzykson C 1971 *Phys. Rev.* **D3** 618–621
[13] Heinzl T *et al.* 2006 *Opt. Commun.* **267** 318–321 (*Preprint* hep-ph/0601076)
[14] Di Piazza A, Hatsagortsyan K Z and Keitel C H 2006 *Phys. Rev. Lett.* **97** 083603 (*Preprint* hep-ph/0602039)
[15] Harvey C, Heinzl T and Ilderton A 2009 *Phys. Rev.* **A79** 063407 (*Preprint* 0903.4151)
[16] Bamber C *et al.* 1999 *Phys. Rev.* **D60** 092004
[17] Berestetskii V, Lifshitz E and Pitaevskii L 1982 *Quantum Electrodynamics (Course of Theoretical Physics, Vol. 4)* (Butterworth-Heinemann, Oxford)
[18] Heinzl T and Ilderton A 2009 *Opt. Commun.* **282** 1879–1883 (*Preprint* 0807.1841)
[19] The HiPER project: http://www.hiper-laser.org/
[20] Jackson J 1999 *Classical Electrodynamics* (Wiley, New York) third edition
[21] Schott G 1912 *Electromagnetic Radiation* (Cambridge University Press)
[22] Landau L and Lifshitz E 1987 *The Classical Theory of Fields (Course of Theoretical Physics, Vol. 2)* (Butterworth-Heinemann, Oxford)
[23] Sengupta N 1949 *Bull. Math. Soc. (Calcutta)* **41** 187–198
[24] Sengupta N 1952 *Bull. Math. Soc. (Calcutta)* **44** 175–180
[25] Kibble T W B 1965 *Phys. Rev.* **138** B740–B753
[26] Volkov D 1935 *Z. Phys.* **94** 250–260
[27] Furry W H 1951 *Phys. Rev.* **81** 115–124
[28] Nikishov A I and Ritus V I 1963 *Zh. Eksp. Teor. Fiz.* **46** 776–796 [Sov. Phys. JETP **19**, 529 (1964)]
[29] Nikishov A I and Ritus V I 1964 *Zh. Eksp. Teor. Fiz.* **46** 1768–1781 [Sov. Phys. JETP **19**, 1191 (1964)]
[30] Narozhnyi N B, Nikishov A and Ritus V 1964 *Zh. Eksp. Teor. Fiz.* **47** 930 [Sov. Phys. JETP **20**, 622 (1965)]
[31] Bula C *et al.* 1996 *Phys. Rev. Lett.* **76** 3116–3119
[32] Chen S, Maksimchuk A and Umstadter D 1998 *Nature* **396** 653
[33] Babzien M *et al.* 2006 *Phys. Rev. Lett.* **96** 054802
[34] Compton A H 1923 *Phys. Rev.* **21** 483–502

Implementation of depolarization due to beam-beam effects in the beam-beam interaction simulation tool GUINEA-PIG++

C Rimbault[1], G Le Meur[1], F Blampuy[1], P Bambade[1] and D Schulte[2]
[1] LAL, Univ Paris-Sud, IN2P3/CNRS, Orsay, France
[2] CERN, Geneva, Switzerland
E-mail: rimbault@lal.in2p3.fr

Abstract. Depolarization is a new feature in the beam-beam simulation tool GUINEA-PIG++ (GP++). The results of this simulation are studied and compared with another beam-beam simulation tool, CAIN, considering different beam parameters for the International Linear Collider (ILC) with a centre-of-mass energy of 500 GeV.

1. Introduction

Precise knowledge of the polarization state of the beams will be required at future linear colliders such as ILC and the Compact LInear Collider (CLIC) [1]. As for many other beam characteristics, the polarization of the beams is modified by beam-beam effects. Such effects are not analytically calculable and beam-beam interaction simulations are thus required.
A treatment of the corresponding spin dynamics has been implemented in GP++ [2] for the beam particles, complementing that already available in the CAIN simulation [3].

2. Depolarization effects during beam-beam interaction

The spin motion in an electromagnetic field is influenced by two effects: the classical spin precession, described by the Thomas-Bargmann-Michel-Telegdi (T-BMT) equation and the spin-flip effect arising from beamstrahlung [4].
The dominant depolarization effect at ILC energies comes from the classical precession of the spin under an electromagnetic field, according to the T-BMT equation:

$$\frac{d\vec{S}}{dt} = -\frac{e}{m\gamma}[(\gamma a + 1)\vec{B_T} + (a+1)\vec{B_L} - \gamma(a + \frac{1}{\gamma+1})\beta\vec{e_v} \times \frac{\vec{E}}{c}] \times \vec{S}. \quad (1)$$

The precession angle is typically γa times the deflection angle, where $a = 0.0011596$ is the anomalous magnetic moment of the electron and γ the Lorentz factor.
The Sokolov-Ternov effect, also called spin-flip, occurs when electrons emit synchrotron radiation. The spin of the radiating electron then has a probability to change its state, proportional to the energy of the emitted photon.

3. Beam spin state after interaction

GP++, as well as CAIN, do not simulate the interaction of the real number of particles, but use macro-particles. Typically, 10^5 macro-particles replace 2×10^{10} real particles. A macro-particle represents thus a statistical ensemble, to which a polarization can be associated as a probability for the spins to be oriented along a given axis. A polarization defined as (0,0,1) is used as initial condition for all the comparisons in this paper, meaning that 100% of the particles have their spin oriented along the beam propagation axis z.
An implementation of the two depolarization mechanisms described above, precession and spin-flip, has been realized in GP++ for the beam macro-particles. The GP++ results are compared to CAIN simulations, for the four beam parameter sets of the ILC at 500 GeV center-of-mass energy (see Table 1)[5].

Table 1. Proposed beam parameters for ILC at 500 GeV center-of-mass energy. N is the number of particles per bunch, $\gamma\epsilon_{x,y}$ respectively horizontal and vertical normalized emittances at the IP, $\beta_{x,y}$ respectively the horizontal and vertical beta functions and $\sigma_{x,y,z}$ respectively the horizontal, vertical and longitudinal sizes of the bunches.

	Nominal	Low N	Large Y	Low P
N [10^{10}]	2	1	2	2
$\gamma\epsilon_x/\gamma\epsilon_y$ [mm.mrad]	10/0.04	10/0.03	10/0.08	10/0.036
$\beta_x/\beta_y/$ [mm]	20/0.4	11/0.2	11/0.6	11/0.2
σ_x/σ_y [nm]	639/5.7	474/3.5	474/9.9	474/3.8
σ_z [μm]	300	200	500	200

3.1. Depolarization from precession

The spin state of the beam is first compared after a full beam-beam interaction, taking into account effects from precession only. As shown in Figure 1, good agreement is obtained between GP++ (red line) and CAIN (blue line) for the distributions of the polarizations of the beam macro-particles in each of the four options. The corresponding mean depolarizations, ΔP, calculated as $1- < P >$,are displayed in Table 2, as well as the deviation between the two simulations, which show that the agreement between GP++ and CAIN results is better than 95%.

Table 2. Depolarization due to precession after a full beam-beam interaction, for the four ILC beam parameter sets at 500 GeV center-of-mass energy.

ΔP [%] from precession	Nominal	Low N	Large Y	Low P
GP++	0.63	0.29	1.39	1.04
CAIN	0.64	0.29	1.46	1.08
(CAIN - GP++)/CAIN [%]	1.56	0	4.79	3.70

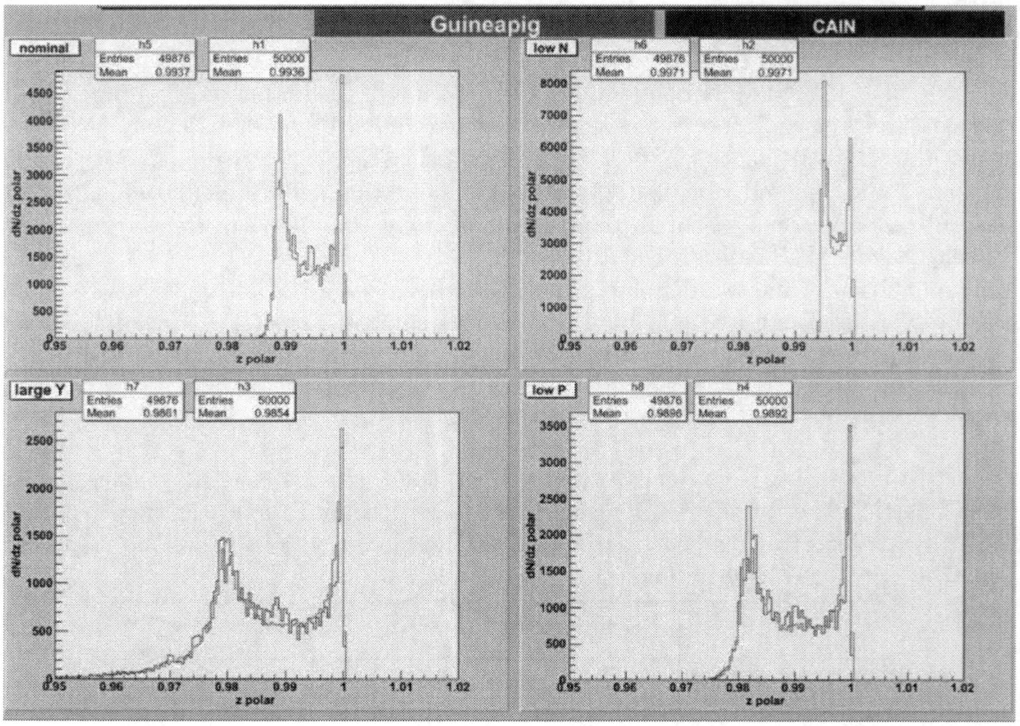

Figure 1. Comparison between GP++ and CAIN simulations for the T-BMT process for the four beam parameter sets of the ILC at 500 GeV center-of-mass energy.

3.2. Total depolarization

Beamstrahlung emission induces an additional spin-flip probability. Figure 2 shows the polarization spectra after the full beam-beam interaction, now taking into account both of the two depolarization effects described above. Mean values are compared in Table 3 . The Sokolov-Ternov mechanism increases the depolarization by about 20% on average for all ILC options.

Figure 3 shows the depolarization as a function of macro-particle energy. One can clearly see the energy-dependance expected from the Sokolov-Ternov spin-flip probability.

Table 3. Total depolarization after a full beam-beam interaction, for the four beam parameter sets of the ILC at 500 GeV center-of-mass energy.

Total ΔP [%]	Nominal	Low N	Large Y	Low P
GP++	0.72	0.36	1.48	1.24
CAIN	0.74	0.36	1.53	1.26
CAIN-GP++/CAIN [%]	2.70	0	3.27	1.59

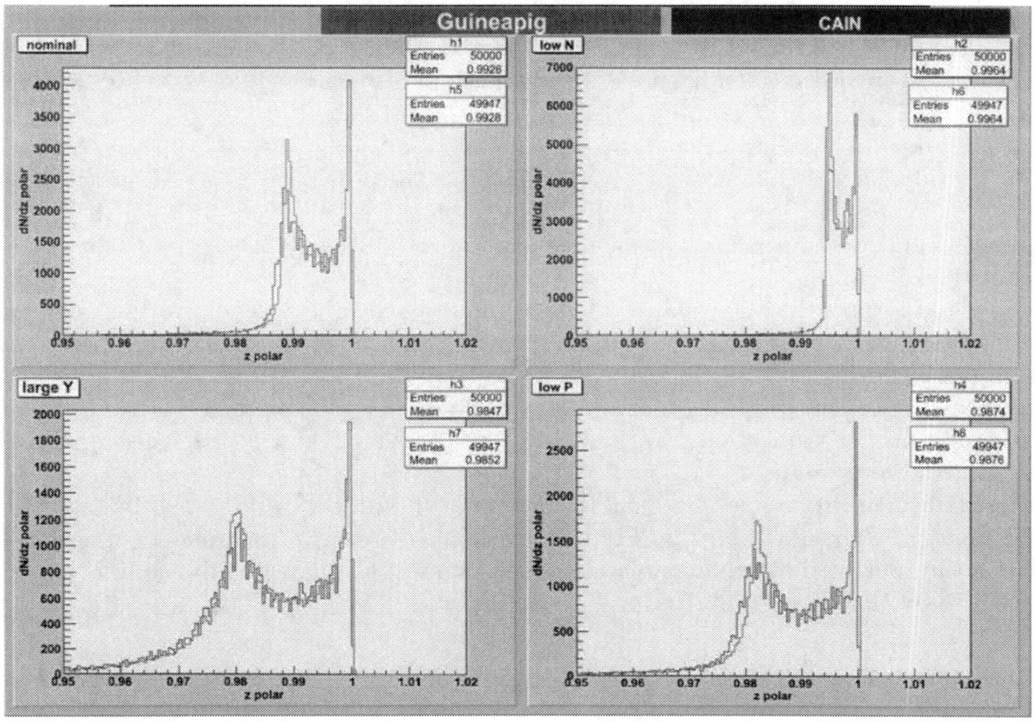

Figure 2. Comparison between GP++ and CAIN simulations for the total depolarization for the four beam parameter sets of the ILC at 500 GeV center-of-mass energy.

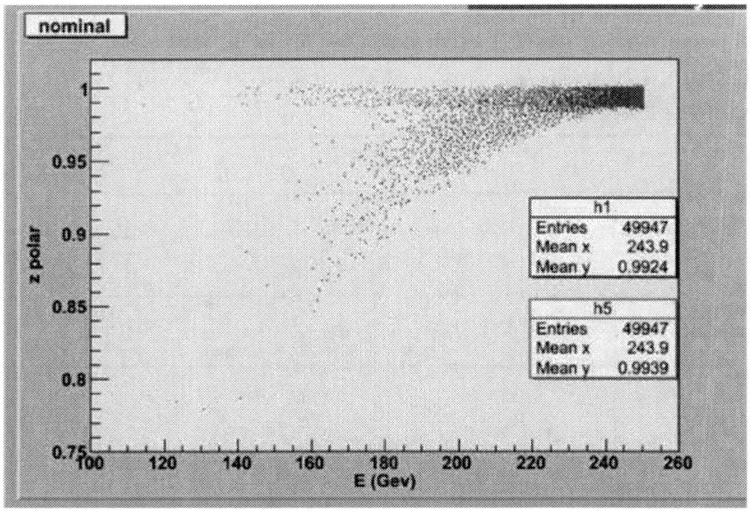

Figure 3. Dependance of the depolarization with the energy of the macro-particles, for the T-BMT precession only (in red) and including also the spin-flip mechanism (in black).

Journal of Physics: Conference Series **198** (2009) 012006 doi:10.1088/1742-6596/198/1/012006

4. Spin state of luminosity contributing beam particles

For physics, it is essential to know the spin state of the particles which interact at the quantum level. This can be obtained by computing the polarization of the beam particles making up the luminosity which is calculated in the simulation, also referred to as luminosity-weighted polarization, P_{lw}.

GP++ provides the polarization state, P_x, P_y, P_z of macro-particles at collision time, whereas CAIN provides specific luminosities corresponding to the possible sets of spin states of the two beams, L_{zz}, L_{zx}, L_{zy}.... Starting from 100% initial polarizations, $P = (0,0,1)$, the luminosity-weighted depolarization, ΔP_{lw}, is calculated as $1 - \sqrt{P_{z1}P_{z2}}$ in GP++, while in CAIN, it is computed as $1 - \sqrt{L_{zz}/L}$.

Tables 4 and 5 show the comparisons for the luminosity-weighted depolarizations computed in the two simulations taking into account, respectively, precession only and both precession and spin-flip mechanisms.

Agreement better than 95% is obtained for the Nominal and Low charge (Low N) ILC beam parameter sets. For the large disruption (Large Y) and Low Power options, discrepancies of up to 20% are however found.

The contribution from spin-flip effects amounts to between 10% and 40% of the total depolarization. Though it can be expected to depend on beam parameters, the 0.273 ratio of luminosity-weighted to total depolarizations which was estimated analytically in [3, 4] is well reproduced by the GP++ simulation.

Table 4. Luminosity-weighted depolarization due to precession for the four 500 GeV center-of-mass energy beam parameter sets of the ILC.

ΔP_{lw} [%] from precession	Nominal	Low N	Large Y	Low P
GP++	0.17	0.08	0.41	0.28
CAIN	0.19	0.09	0.48	0.30
CAIN-GP++/CAIN [%]	10	11	15	7

Table 5. Total luminosity-weighted depolarization for the four 500 GeV center-of-mass energy beam parameter sets of the ILC.

Total ΔP_{lw} [%]	Nominal	Low N	Large Y	Low P
GP++	0.23	0.13	0.46	0.41
CAIN	0.24	0.13	0.57	0.50
CAIN-GP++/CAIN [%]	4	0	19	18

5. Dependance with beam parameters

The dependance of the luminosity-weighted depolarization with beam parameter variations was also tested with GP++.

Journal of Physics: Conference Series **198** (2009) 012006 doi:10.1088/1742-6596/198/1/012006

As displayed in Figure 4, no drastic dependances are observed for horizontal or vertical beam offsets in ranges for which a reasonable luminosity is maintained. For a vertical beam offset of two sigma, which is very pessimistic, an absolute variation of only 0.03% is found for the luminosity-weighted depolarization. The dependance with vertical beam size also shows good stability for variations of up to 50%.
There is however a large sensitivity to changes in the horizontal beam size. In Figure 5 it is shown that a 10% variation would lead to an absolute uncertainty of 0.05% on the depolarization, representing a 20% relative error.

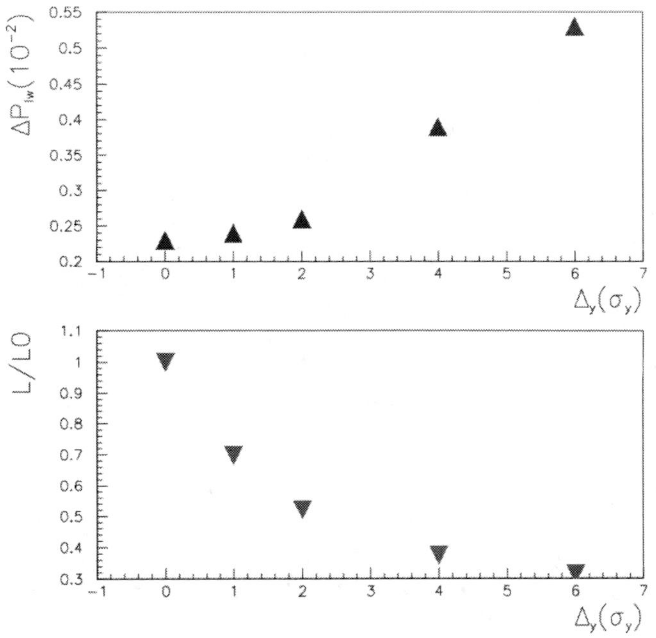

Figure 4. Dependance of the luminosity-weighted depolarization (upper) and of the luminosity (lower) on the vertical offset for the Nominal ILC beam parameter set at 500 GeV center-of-mass energy.

6. Conclusion

A spin dynamics treatment including precession and spin-flip effects has been implemented in GP++, following that already available in the CAIN simulation. Studies of the depolarizing effects induced by the beam-beam interaction using this new simulation show both reasonable qualitative and quantitative behaviour.
Explicit comparisons with the CAIN simulation are presently the only way to test this new development in GP++. Although some differences are found, good overall agreement is obtained for the main ILC beam parameter sets.
For Nominal ILC parameters at 500 GeV, a luminosity-weighted depolarization of $0.23 \pm 0.01\%$ is predicted with both simulations, confirming prior studies using CAIN[6].
A first study of the dependance with beam parameters indicates that the horizontal beam size must be controlled to better than 10% to maintain a relative precision of better than 20% on the degree of polarization.

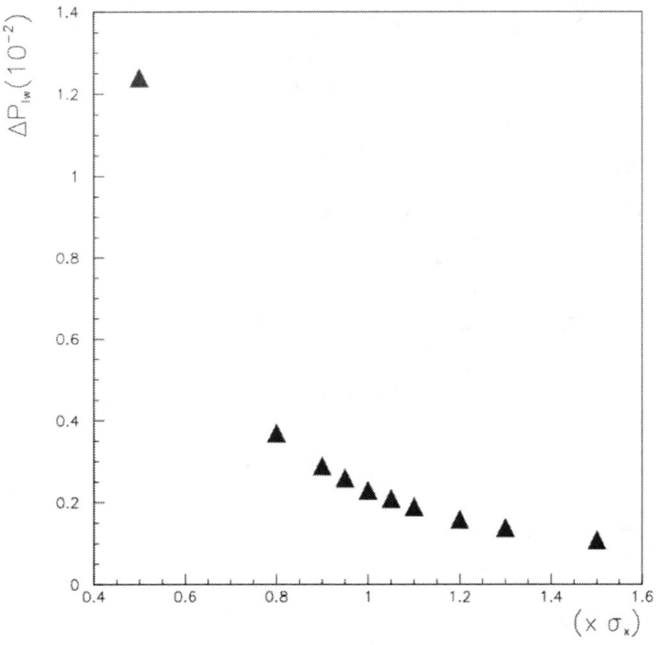

Figure 5. Dependance of the luminosity-weighted depolarization on the horizontal beam size for the Nominal ILC beam parameter set at 500 GeV center-of-mass energy.

7. Acknowledgement

We would like to thank Ian Bailey, Anthony Hartin and Gudrid Moortgat-Pick for useful discussions. This work is supported by the Commission of the European Communities under the 6th Framework Programme "Structuring the European Research Area", contract number RIDS-011899.

References

[1] A Djouadi, J Lykken, K Mönig, Y Okada, M Oreglia and S Yamashita, "The Physics Case for the International Linear Collider", Frozen draft for the ACFA meeting in Beijing, February 2007 http://www-zeuthen.desy.de/ILC/dcr_physics/dcr_beijing.pdf
[2] G Le Meur et al., EUROTeV-Report-2008-067, 2008 http://flc.web.lal.in2p3.fr/mdi/BBSIM/bbsim.html
[3] K Yokoya, User's Manual of CAIN - Version 2.3, october 2001 http://www-acc-theory.kek.jp/members/cain/default.html
[4] K Yokoya and P Chen, SLAC-PUB-4692, 1988
[5] International Linear Collider Reference Design Report, first draft, february 2007 http://media.linearcollider.org/rdr_draft_v1.pdf
[6] I R Bailey et al., EUROTeV-Report-2008-026, 2008

Joint IPPP, CI and ICFA Workshop on Advanced QED Methods for Future Accelerators IOP Publishing
Journal of Physics: Conference Series **198** (2009) 012007 doi:10.1088/1742-6596/198/1/012007

Does experiment show that beamstrahlung theory - strong field QED - can be trusted?

J. Esberg, U.I. Uggerhøj

Department of Physics and Astronomy, University of Aarhus, Denmark

Abstract. We present a discussion of the experimental foundation for the strong field effects that are applied in beamstrahlung calculations for the planned next generation of linear colliders. In particular, quantum suppression of synchrotron radiation, coherent pairs and tridents are treated. It is shown, that many of these phenomena can be - and some have been - tested using present day technology.

1. Introduction

The admirable successes of precision tests of Quantum ElectroDynamic (QED) processes, where experiment and theory have been shown to agree with an accuracy down to 10^{-14} - for instance in the hydrogen $1S - 2S$ transition or the $g - 2$ experiments [1] - has naturally led to a high degree of confidence in extrapolations of QED to hitherto untested regimes. New experiments, demanding extreme precision, as e.g. the $1S - 2S$ transition comparison between hydrogen and antihydrogen which aim at an ultimate accuracy approaching 10^{-18}, see e.g. [2], are for at least the first 12 orders of magnitude - corresponding to the foreseeable future - purely QED experiments. Therefore, and since physics is basically an experimental science, it still makes sense to ask the question: Is QED really correct in new regimes, and to what extent can we - based on experimental evidence - have confidence in QED, e.g. in the strong field regime?

In the following, we present a number of measurements that have been performed using strong crystalline fields in order to access the QED regime in which one of the three relativistic invariants that can be formed from the electromagnetic field tensor and the four-momentum of the particle (electron or photon) is comparable to, and in some cases even substantially exceeds, one. The strong-field regime is where the field in the rest frame of the electron becomes comparable to the so-called critical field - in QED the only combination of the electron mass and charge, Planck's constant and the speed of light that yields an electric field, signifying that relativistic quantum field theory must be applied. This regime is interesting in its own right, showing clear quantum corrections to a theory that is to a large extent describable by purely classical concepts [3]. But it is also of crucial importance to the construction of the next generation of linear colliders. In these machines, the electromagnetic field of one bunch acts upon a particle in the opposing bunch in exactly this manner: a critical field is experienced by the electron. This leads to a new behaviour of the synchrotron radiation equivalent of particle deflection in the field of the bunch, instead of in a magnetic dipole: Beamstrahlung. As the emission of beamstrahlung has a direct and significant impact on the energy of the colliding particles, it is a decisive factor for e.g. the energy-weighted luminosity. It is therefore important to know if beamstrahlung theory

Journal of Physics: Conference Series **198** (2009) 012007 doi:10.1088/1742-6596/198/1/012007

is correct for the conceptual and technical design of the collision region - the center about which the rest of the machine is based.

Three dimensionless invariants can be constructed from the the electromagnetic field strength tensor, $F_{\mu\nu}$, and the momentum four-vector p^ν (or, in the case of a photon, $\hbar k^\nu$). Two of these, $\Xi = F_{\mu\nu}^2/\mathcal{E}_0^2 = 2(\vec{B}^2 - \vec{\mathcal{E}}^2)/\mathcal{E}_0^2$ and $\Gamma = e_{\lambda\mu\nu\rho}F^{\lambda\mu}F^{\nu\rho}/\mathcal{E}_0^2 = 8\vec{\mathcal{E}}\cdot\vec{B}/\mathcal{E}_0^2$ express the 'inherent' field strength and 'nature' (pure electric or magnetic in some frame) of the field, respectively, taken in units of the critical field $\mathcal{E}_0 = m^2c^3/\hbar e$. The third is given by

$$\chi^2 = \frac{(F_{\mu\nu}p^\nu)^2}{m^2c^4\mathcal{E}_0^2} \tag{1}$$

expressing the field experienced by the particle. For an ultra-relativistic particle moving across fields $\mathcal{E} \ll \mathcal{E}_0, B \ll B_0$ with an angle to the field direction $\theta \gg 1/\gamma$ the invariants fulfill $\chi \gg \Xi, \Gamma$ and $\Xi, \Gamma \ll 1$. The relation of χ to the fields $\vec{\mathcal{E}}$ and $\vec{B}$ is given by [4]

$$\chi^2 = \frac{1}{\mathcal{E}_0^2 m^2 c^4}((c\vec{p}\times\vec{B} + E\cdot\vec{\mathcal{E}})^2 - (c\vec{p}\cdot\vec{\mathcal{E}})^2) \tag{2}$$

For an ultrarelativistic particle moving perpendicularly to a pure electric or pure magnetic field this reduces to

$$\chi = \frac{\gamma\mathcal{E}}{\mathcal{E}_0} \quad \text{or} \quad \chi = \frac{\gamma B}{B_0} \tag{3}$$

Thus, a Lorentz transformation reveals that electromagnetic fields that in the restframe are transversal to the boost remain transversal and are larger by a factor of γ. This is a consequence of the relativistic invariance of the parameter χ in equation (3). The critical field $\mathcal{E}_0$ is frequently referred to as the Schwinger field [5], although it was treated as early as 1931 by Sauter [6, 7], following a supposition by Bohr. The χ parameter - usually called Υ in the accelerator physics community - is decisive for the behaviour of the emission of beamstrahlung in 0.5 TeV linear colliders. The question is: Can processes involving $\chi \simeq 1$ be addressed experimentally with present day technology, such that estimates of the reliability of QED in beamstrahlung for future machines can be given now? In the following, we answer this question in the affirmative.

2. Quantum synchrotron

As the field in a synchrotron radiation experiment grows larger, the recoil imposed on the electron by emitting increasingly energetic photons becomes more and more important. If this correction is not taken into account, the calculated synchrotron radiation will carry away more energy than the energy of the charged particle - clearly an impossible situation. This enforces an inclusion of the quantum nature of the radiation process in the description of synchrotron radiation in strong fields. The replacement in the classical formula of synchrotron radiation of the frequency ω with $\omega^\star = \omega E/(E - \hbar\omega)$ provides an excellent description of the spinless quantum regime, i.e. radiation from a charged scalar particle [8]. However, for $\chi \gg 1$ the hard end of the photon spectrum becomes dominated by a spin flip transition, making it develop a pronounced peak near the incident electron energy. This feature allows the transfer of helicity in strong fields at large photon energies [9]. Meanwhile, electron motion in a transversal electric field generates linearly polarized light in non spin-flip transitions.

2.1. Quantum treatment of beamstrahlung

The discussion of quantum effects - and in particular a suppression of the intensity - in radiation emission from energetic particles in collision with a counterpropagating bunch, was started in the mid-80's [10, 11, 12] and later continued with treatments of pair creation [13, 14]. It turned

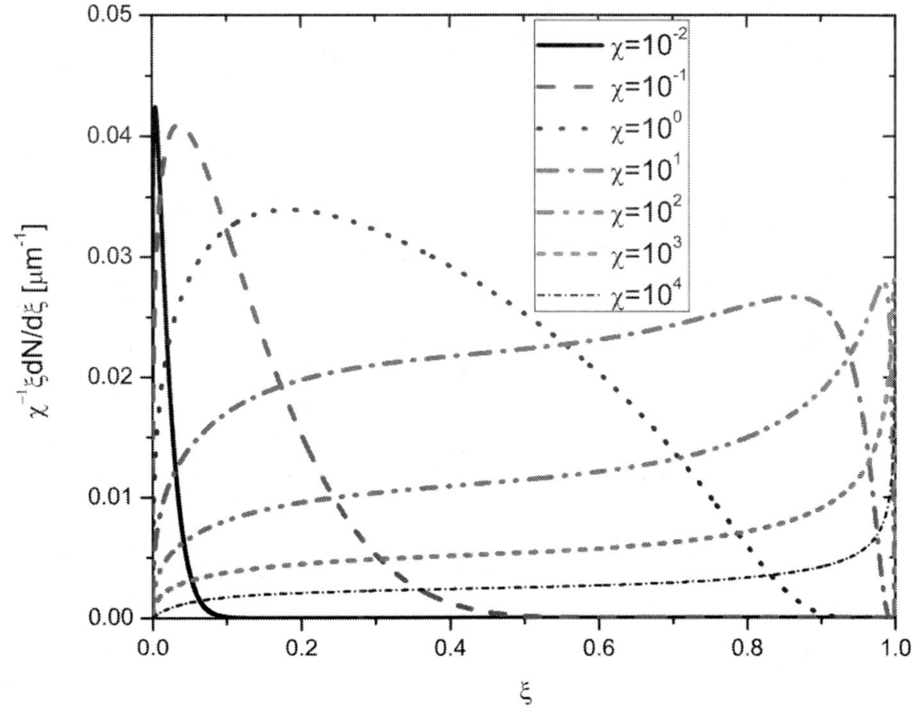

Figure 1. *Synchrotron spectrum dependence on the magnitude of χ. For small χ values the spectrum tends to a classical synchrotron spectrum. For $\chi \ll 1$ the spectrum, when scaled to the Lienard energy loss and the critical frequency, is invariant [3]. On the other hand, for χ0.1, the spectrum shape becomes strongly dependent on the exact value of χ and for $\chi \gg 1$ peaked towards the energy of the incoming particle, i.e. typically $\xi = \hbar\omega/E_0 \rightarrow 1$.*

out, that the emission of synchrotron-like radiation from the beams enters the regime where χ becomes of the order one, and the spectrum changes character completely, see figure 1. Of particular relevance to the connection of beamstrahlung with emission from electrons penetrating crystals is the contribution by Baier and Katkov - two of the theory pioneers in strong field effects in crystals - to the early development of the theory of beamstrahlung [15].

Shortly after the first publications on the relevance of quantum theory to beamstrahlung, Blankenbecler and Drell contributed a full quantum treatment of the problem, based on the eikonal approximation [16]. The scaling parameter in this case is given by

$$C = \frac{m^2c^3RL}{4Ne^2\gamma^2\hbar} \tag{4}$$

representing the electric field from a homogeneously charged cylinder of length L and radius R holding N charges, in units of $\mathcal{E}_0 = m^2c^3/\hbar e$. The applicability of this scaling parameter was later elaborated upon by Solov'yov and Schäfer [17, 18]. From this model of a beam, the form factor $F = \delta/\delta_{\text{classical}}$, describing the quantal energy loss in units of the classical, was derived and approximated by:

$$F(C) = [1 + \frac{1}{b_1}[C^{-4/3} + 2C^{-2/3}(1 + 0.20C)^{-1/3}]]^{-1} \tag{5}$$

with $b_1 = 0.83$, see also [17]. Clearly, as stated by Blankenbecler and Drell, in the classical

Journal of Physics: Conference Series **198** (2009) 012007 doi:10.1088/1742-6596/198/1/012007

regime $\hbar \to 0$ in equation (4) such that C tends to infinity, and therefore the form factor tends to 1 according to equation (5), as must be required.

Even earlier, Himel and Siegrist [10], developed a fairly compact expression for the fraction of the energy radiated by the electron

$$\delta_{QM} = \frac{8mc^2}{\sqrt{3}}\left(\frac{r_e}{2\sqrt{3}\hbar c}\right)^{4/3}\left(\frac{DP}{\gamma f}\right)^{1/3} \tag{6}$$

where $r_e = \alpha\lambda$ is the classical electron radius, $D = Nr_e\sigma_z/2\gamma\sigma_r^2$ is the disruption parameter for which σ expresses the beam radius, N the number of particles per bunch, $P = fN\gamma mc^2$ the beam power and f the repetition rate.

Blankenbecler and Drell in their paper state "'Their [Himel and Siegrist's [10]] final formula [eq. (6)] is remarkably accurate in the full quantum regime: $C \ll 1$."' It is remarkable, but not surprising, as several thorough investigations have since shown, e.g. [19, 15] and as we will now show with relatively simple means. The approach of Himel and Siegrist rests on an estimate based on the "'... full quantum treatment of synchrotron radiation [which] was done in 1952 by Sokolov, Ternov and Klepikov [20]"', and also by Julian Schwinger [21].

As a result of the quantum correction, the total radiated intensity for the classical emission is according to Schwinger reduced by a factor $1 - 55\sqrt{3}\lambda_c\omega_0\gamma^2/16c$ due to first order quantum corrections when $\chi \ll 1$ [20, 21, 22]. Including the second order term the reductions for small and large values of χ are [4, 23]

$$I/I_{\rm cl} = 1 - 55\sqrt{3}\chi/16 + 48\chi^2 \qquad \chi \ll 1 \tag{7}$$

$$I/I_{\rm cl} \simeq 0.56\chi^{-4/3} \qquad \chi \gg 1 \tag{8}$$

Furthermore, an approximate expression ("accuracy better than 2% for arbitrary χ" [23, eq. (4.57)])

$$I/I_{\rm cl} \simeq (1 + 4.8(1+\chi)\ln(1+1.7\chi) + 2.44\chi^2)^{-2/3} \tag{9}$$

gives a compact analytical expression applicable e.g. in computer codes.

In figure 2 is shown graphs based on equation (5) and equation (9), where it has been assumed that $C = 1/\chi$. Clearly the curves are similar, and the likelihood of the formulas originating from the same phenomenon becomes even more evident by adjusting to $C = 1.3/\chi$ which results in the curves being indistinguishable on the plot. This is remarkable, given a range of more than five orders of magnitude in χ. In Blankenbecler and Drell's theory, the bunch is treated as a homogeneously charged cylinder of length L and radius R holding N charges. At the distance r from the center of this cylinder the electric field is $2\gamma Ne/Lr$, which leads to an average (over r) field

$$\mathcal{E} = \frac{\gamma Ne}{LR} \tag{10}$$

which can be combined with equation (4) to give

$$C = \frac{m^2c^3}{\gamma\mathcal{E}e\hbar} = \frac{\mathcal{E}_0}{\gamma\mathcal{E}} = \frac{1}{\chi} \tag{11}$$

so it is legitimate to interchange C and $1/\chi$. The reason for the additional factor 1.3 that brings the curves into almost exact agreement, is due to the radiation intensity being non-linear in C, i.e. averaging over the field encountered and then calculating the intensity from this field is not the same as calculating the intensity from the fields encountered and then averaging.

In addition, early studies by Chen and Yokoya [24] showed that field gradient effects are small for a collider operating near $\Upsilon = \chi = 1/C = 1$, i.e. also for the planned Compact LInear Collider (CLIC) at CERN where the expected value is $\Upsilon \simeq 4$. It is therefore to a high degree of accuracy sufficient to use equation (5) derived for the homogeneously charged cylinder.

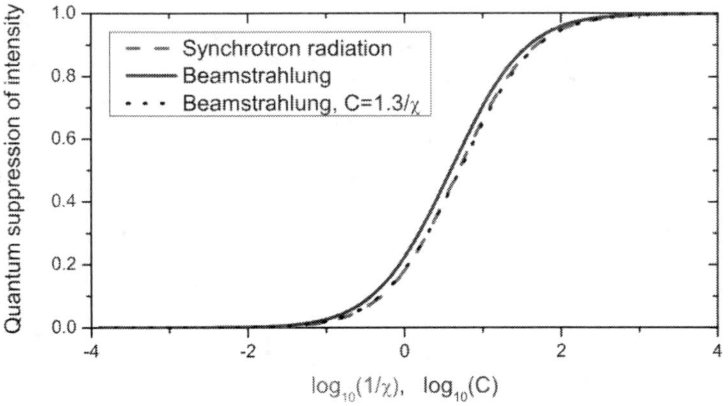

Figure 2. *The quantum suppression of radiation emission intensity, according to eq. (9) and eq. (5).*

2.2. Experiments that support the theory

With the acceleration techniques at hand today, the energy of electrons available for laboratory experiments is limited to $\gamma \leq 10^6$. This means that *macroscopic* magnetic fields of the order a few thousand Tesla or electric fields of the order $\mathcal{E} = 10^{11}$ V/cm are needed to experimentally test the theory for $\chi = \gamma\mathcal{E}$ [V/cm]/$1.32 \cdot 10^{16} \simeq 1$. At present this is certainly beyond reach by conventional means. However, an analogous case is seen for the pair production and radiation emission in single crystals and these phenomena enable a verification of the essential expressions in beamstrahlung theory.

In figure 3 is shown results for radiation emission from electrons impinging on a tungsten crystal close to the $\langle 111 \rangle$ axis. As a consequence of the strong deflection upon the passage of the string of nuclei composing the axis, the electron is forced to emit radiation as in a constant field. This happens much like in normal synchrotron radiation emission, only in a much more intense field, 10^{11} V/cm, corresponding to 30.000 T. Put in one sentence only, such fields are obtained from the coherent addition of the nuclear electric fields in the crystal - fields that for unscreened nuclei of high Z are themselves only slightly smaller than $\mathcal{E}_0$ at a distance of a 1s electron, $\mathcal{E}_{1s}/\mathcal{E}_0 = \alpha^3 Z^3$. As a result of the high peak value of the χ parameter ($\chi_{W,\langle 111 \rangle} \simeq 0.03 \cdot E$[GeV]), the radiation emission is subjected to strong quantum suppression. For a brief introduction to these effects in crystals, see Appendix A. In the limit $\chi \ll 1$ the enhancement - defined as the intensity from the aligned crystal divided by the intensity for a non-aligned (so-called 'random' i.e. effectively amorphous) crystal - would be linear with increasing energy, as shown by the dash-dotted line. This is the case because synchrotron radiation emission is quadratic in energy and radiation from an amorphous foil is linear in energy, but due to the strong quantum suppression, the enhancement is reduced to the level shown by the dashed line, as also expected from equation (9). The good agreement between experimental values and theory shown in figure 3, combined with the equality of beamstrahlung and strong field theory shown in figure 2 provide a strong experimental indication that QED theory as applied to beamstrahlung in the regime 1Υ10 is correct.

The accuracy of the experimental values is $5 - 10\%$, enough to ascertain the validity of

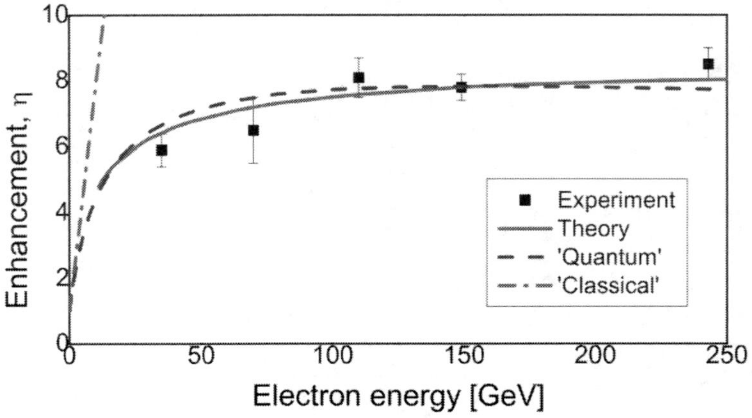

Figure 3. *Experiment [25] and theory [26] for radiation emission from electrons penetrating a tungsten crystal near the ⟨111⟩ axis. For comparison, a curve based on equation (9) with a slightly arbitrary, but realistic* $\overline{\chi} = 0.02 \cdot E[GeV]$ *(and vertical scale obtained as the best fit) is shown as the dashed line ('quantum'), and the corresponding classical expression as the dash-dotted line ('classical'). The enormous difference between the 'classical' and 'quantum' curves directly show the strong quantum suppression in the experimentally accessible regime.*

the theoretical approach. We can therefore confidently claim that the 'quantum synchrotron' behaviour of radiation emission in a strong field, is experimentally well confirmed.

3. 'Coherent pairs'

Having now established the equality between the variables $1/C$ and Υ used in the beamstrahlung community and the variable χ used to describe synchrotron radiation in a strong field, we proceed with a discussion of pair creation in strong fields, described, as expected from crossing symmetry of the processes, by similar variables.

3.1. Pair production in a strong electromagnetic field

In the following, pair production and radiation emission processes are considered as taking place in magnetic or electric fields that are homogenous (constant in space over the formation length of the corresponding process). The pair production probability differential in the energy of one of the final state particles, ϵ, is given as [23]

$$\frac{\mathrm{d}W}{\mathrm{d}\epsilon} = \frac{\alpha m^2 c^4}{\sqrt{3}\pi\hbar^3\omega^2}[\frac{\epsilon^2+\epsilon_f^2}{\epsilon\epsilon_f}K_{2/3}(\xi) + \int_\xi^\infty K_{1/3}(y)\mathrm{d}y] \tag{12}$$

where $\epsilon_f = \hbar\omega - \epsilon$, $\xi = 8u/3\chi_p$, $u = \gamma^2$, and $\mathrm{d}W = \mathrm{d}w/\mathrm{d}t$ is the probability per unit time. Here ϵ denotes the energy of one of the produced particles, ϵ_f that of the other and $\hbar\omega$ the energy of the photon. The functions $K_{1/3}$ and $K_{2/3}$ are the modified Bessel functions of order $1/3$ and $2/3$, respectively. The Lorentz invariant pair production strong field parameter is given as $\chi_p = \gamma_p B/B_0$ where the critical field once again appears. γ_p is understood here as $\hbar k_\perp/mc$, i.e. the photon momentum transverse to the magnetic field, in units of mc. The 'strong field regime' is reached when χ_p becomes comparable to or exceeds 1 - but already at $\chi_p \simeq 0.1$ strong field effects become significant.

The differential probability develops a pronounced minimum around $x = \epsilon/\hbar\omega = 1/2$ and peaks at $x \simeq 1.6/\chi_p$ for large χ_p.

The total pair production probability per unit time is given as [23]

$$W = \frac{\alpha m^2 c^4}{6\sqrt{3}\pi\hbar^2\omega} \int_1^\infty \frac{8u+1}{u^{3/2}\sqrt{u-1}} K_{2/3}(\xi)\mathrm{d}u \tag{13}$$

In the limit $\chi_p \ll 1$ the probability is exponentially small, $W \simeq 3\sqrt{3}\alpha m^2c^4\chi_p/16\sqrt{2}\hbar^2\omega \exp(8/3\chi_p)$, and in the limit $\chi_p \gg 1$ the result is $W \simeq 0.38\alpha mc^2 B/B_0\hbar\chi_p^{1/3}$, i.e. the probability actually diminishes once the energy surpasses the domain where $\chi_p \simeq 10$ [23].

The equations (12) and (13) are precise expressions (for $B \ll B_0$), but eq. (13) has an 'analytic approximation' that does not involve integration [27] (where $\chi_e = \hbar\omega B/2mc^2B_0$)

$$n_{\rm pairs} \simeq n_{\rm photons} d\frac{\alpha B}{2\lambda B_0} \cdot 0.16\frac{1}{\chi_e}K_{1/3}^2(2/3\chi_e) \tag{14}$$

and which generally forms the basis for many calculations.

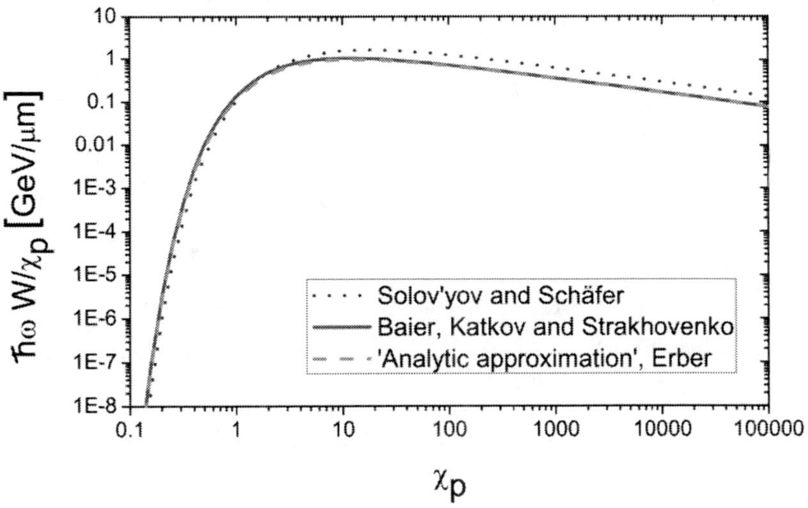

Figure 4. *Comparison between the total pair production probabilities in strong fields of colliding beams according to Solov'yov and Schäfer [18] (dotted line), in a general strong field according to Baier, Katkov and Strakhovenko [23] (full line) and the 'analytic approximation' to the latter, eq. (14), from Erber [27] (dashed line).*

In an article by Solov'yov and Schäfer [18] pair creation by photons colliding with a relativistic beam is treated. Once again, their central scaling parameter

$$C_p = \frac{m^3c^5 r}{2e\mathcal{E}\gamma\omega R} \tag{15}$$

is closely related to χ_p by $C_p = r/\chi_p R$ where R is the radius of the beam colliding with the photons and r is the position of the photon with respect to the beam. In their article an averaging over r is done, but naively assuming $r = R$ one can compare directly with the results of Baier and Katkov, cf. equation ((13)). This simple comparison, shown in figure 4, shows that the functional dependence on the strong field parameter of pair creation is the same.

3.2. Experiments that support the theory

In crystals, the process of axial pair creation is well described by the strong field equations mentioned above. Multiple scattering events give rise to an LPM (Landau-Pomeranchuk-Migdal) effect [28] which is a reduction of the order of a few percent to the total pair production rate and especially dominant for soft photons. Furthermore there is an incoherent contribution to the pair production rate which dominates the spectrum at low photon energies, while this contribution is nearly negligible compared to the total rate when the photon energy is large. Pair production probability rises above the usual Bethe-Heitler one with increasing photon energy. This is a consequence of the strong field and has been measured in several crystalline media [29].

The experimental results are well understood theoretically, and though the constant field approximation does well in estimating the yield, a more careful theoretical treatment is needed accounting for the incoherent contribution as well as the LPM effect and atomic scattering nature of the process. Again QED theory proves sufficient for description of the coherent pairs in crystalline matter, see figure 5.

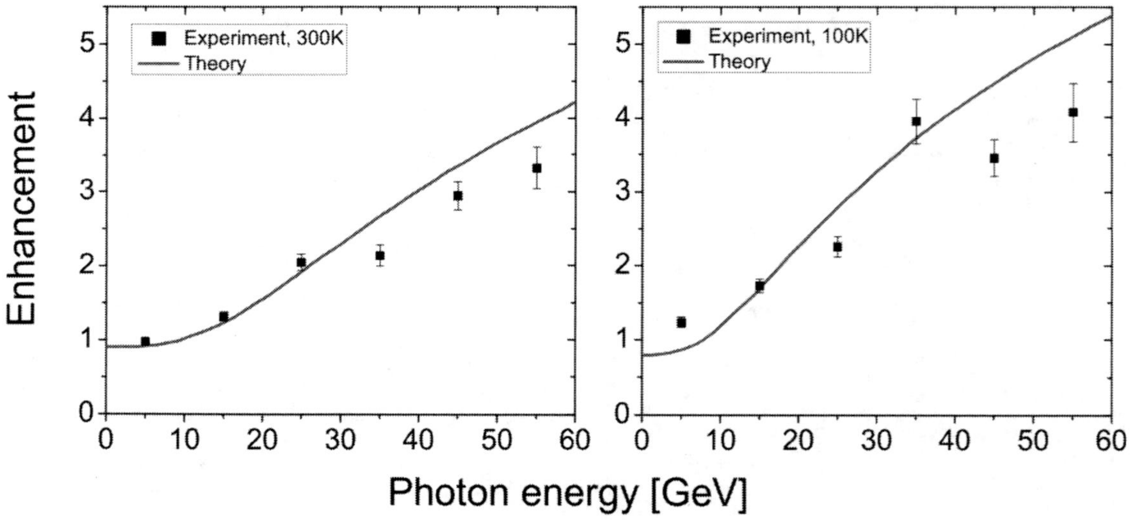

Figure 5. *Total pair production enhancement in W ⟨111⟩ as function of photon energy. Theoretical curves from [28] and experiment from [29]. The coherence rises due to reduced thermal vibrations, making cooling of the crystals a way of pronouncing the strong field effect.*

4. 'Landau-Lifshitz' pairs and tridents

In a strong field, electron/positron pairs may be produced via a virtual photon. The direct production of a pair via virtual photons of a constant field can be calculated by the Weizsäcker-Williams method of virtual quanta. In this approach, the spectral density of virtual photons carried by an electron of energy E_0 may be estimated by

$$\frac{dN_{ww}}{\hbar\omega} = \frac{2}{\pi}\frac{\alpha}{\hbar\omega}W\left(\frac{\hbar\omega}{E_0}\right), \tag{16}$$

where

$$W(x) = xK_0(x)K_1(x) - \frac{x^2}{2}(K_1^2(x) - K_0^2(x)) \tag{17}$$

The amount of these photons converting in a strong field corresponds to the number of Landau-Lifschitz pairs. The subject of equivalent-quanta trident production in a homogenous field has been studied by several authors [30, 27] or in crystals using the Weizsäcker-Williams method [31]. Since no electromagnetic quanta need to be created, the dependence on the thickness of the crystal is linear. The total probability of this process in a homogenous field is [30]

$$W_v = \frac{13\alpha^2}{18\sqrt{3}\pi\lambda_c}\frac{\chi mc^2 t}{E_0}(\ln(\chi) - 0.577 - \ln(4\sqrt{3}) - 77/39) \tag{18}$$

where t is the extent of the magnetic field. The characteristic length of formation for this process is $l_v = \lambda B_0/B$.

A competing process of a bremsstrahlung photon being emitted and the subsequent conversion of the photon into a pair occurs, but it has a quadratic dependence on the thickness. The combined virtual (direct) and real (cascade) photon quanta processes have become known as the 'trident' process [27] due to the shape of the tracks in emulsions for this type of event. The total probability of cascade tridents is within 4% accuracy for arbitrary χ [30] found by the expression

$$W_r = \frac{3\chi}{16}\left(\frac{\alpha t}{\gamma\lambda_c}\right)^2 \ln(1 + \frac{\chi}{12})\exp(-\frac{16}{3\chi})(1 + 0.56\chi + 0.013\chi^2)^{1/6}, \tag{19}$$

where λ_c is the reduced Compton wavelength, $\lambda_c/2\pi$. The formation length measured in the laboratory is given for soft photons as

$$l_f = \frac{2\gamma^2 c}{\omega} \tag{20}$$

this is the characteristic length for radiation emission in a classical synchrotron experiment. It may be seen as the distance an electron needs to travel before the created photon is separated by one reduced wavelength. By choosing an extent of the magnetic field comparable to the formation length, the cascade contribution to the trident yield will be suppressed since the formation length of the cascade process is larger than the formation length for a direct trident.

In beam-beam collisions, it is crucial to know the yield of the direct process since the dependence on χ is very strong for relevant values of χ, and the lack of a photon track may make it a difficult background to account for.

A recent experiment [32] addressed trident production using aligned Ge $\langle 110\rangle$ crystals. Results are still preliminary, but a strong field effect on the trident process was seen. The trident crystal-to-amorphous enhancement shows that crystals can be used to observe higher order QED processes in strong fields. Theoretical work on the issue [33] has recently been published suggesting that the dominant process is the cascade process for the configuration investigated, meaning that the enhancement of the trident yield is closely connected to the enhancement of bremsstrahlung photons.

5. Spin-flip in a strong field

In the early 90's, Lindhard[8] showed that the contribution from the spin can be derived from a Weizsäcker-Williams type calculation, replacing the spin-less Thomson cross section by the

Klein-Nishina cross section. He and Sørensen [34] convincingly demonstrated that the contribution from the spin dominates the hard end of the photon spectrum as soon as $\chi \gg 1$. This means that - apart from the reverse action of the photon on the electron, the recoil - an additional quantum effect of the spin of the electron influences the spectrum.

Why the spin makes its influence can be seen by a rough, but simple argument [35]: The energy of a magnetic moment $\vec{\mu}$ at rest in a magnetic field $\vec{B}$ is given by

$$E_\mu = -\vec{\mu} \cdot \vec{B} \tag{21}$$

Therefore spin-flip transitions of electrons with $\mu = e\hbar/2mc$ have an energy $\Delta E_\mu = e\hbar B/mc$ in the rest system. In this system, the purely electric field in the crystal is by the Lorentz transformation converted into a magnetic field $B = \gamma\beta\mathcal{E}_{\mathrm{lab}}$ plus an electric field which does not influence $\vec{\mu}$. Transformation back to the laboratory yields a factor γ (as in the case of channeling radiation arising from transitions in the transverse potential) such that the result is

$$\Delta E_\mu = \gamma^2 \beta \frac{\mathcal{E}}{\mathcal{E}_0} mc^2 = \chi\beta\gamma mc^2 \tag{22}$$

which is equal to the initial energy of the electron, E, when $\chi\beta = 1$. This simple estimate shows why the radiation from spin-flip concentrates near the endpoint of the spectrum. Asymptotically, the spin contribution becomes $\xi dN/d\xi \propto (\xi^7/(1-\xi))^{1/3} \cdot \chi^{2/3}$ for $\chi \to \infty$ such that it is strongly peaked at the upper end of the spectrum, see figure 1.

Considering the process of radiative polarization of the electron, the typical time of spin-flip transitions, τ, it is given by [22]

$$\tau = \frac{8\hbar}{5\sqrt{3}\alpha m} (\frac{B_0}{B})^3 \frac{1}{\gamma^2} = \frac{8\hbar}{5\sqrt{3}\alpha m} \frac{\gamma}{\chi^3} \tag{23}$$

such that $c\tau$ becomes 15 μm for a 150 GeV electron in a $\chi = 1$ field. Therefore a substantial fraction of the radiation events originate from spin-flip transitions as one gets to and beyond $\chi \simeq 1$, even in a target as thin as 0.1 mm.

5.1. Experiments that support the theory

In figure 6 is shown the radiated power as a function of the fractional photon energy $\xi = \hbar\omega/E$ for 2 energies of the impinging electron, 70 and 243 GeV (more examples can be found in [35]). The experimental points are compared to two theoretical curves: one where the effect of spin has been neglected and one where the electron is treated properly as a spin $^1/_2$ particle. Although the experiment suffers a bit from lack of sufficient statistics (given indirectly by the scatter of points), the trend is clear: The spin of the electron must be included to get the best description of the experimental values. This is not surprising, but it shows that the above description of spin-flip processes is likely to be correct, even in a strong field. It must be emphasized, though, that neither the electron nor the photon beam becomes polarized in the experiment - the electron is exposed to strong fields, but in the radial direction from the crystal axis. Since this does not single out any preferred direction, the beams cannot become polarized, yet the emission spectrum may bear evidence for the spin-flip process.

6. Related areas where similar theories are applied

Strong field effects can be investigated by other means. One example is in heavy ion collisions where the field becomes comparable to the critical field, but the collision is of extremely short duration. Another - technically demanding - example is in multi-GeV electron collisions with

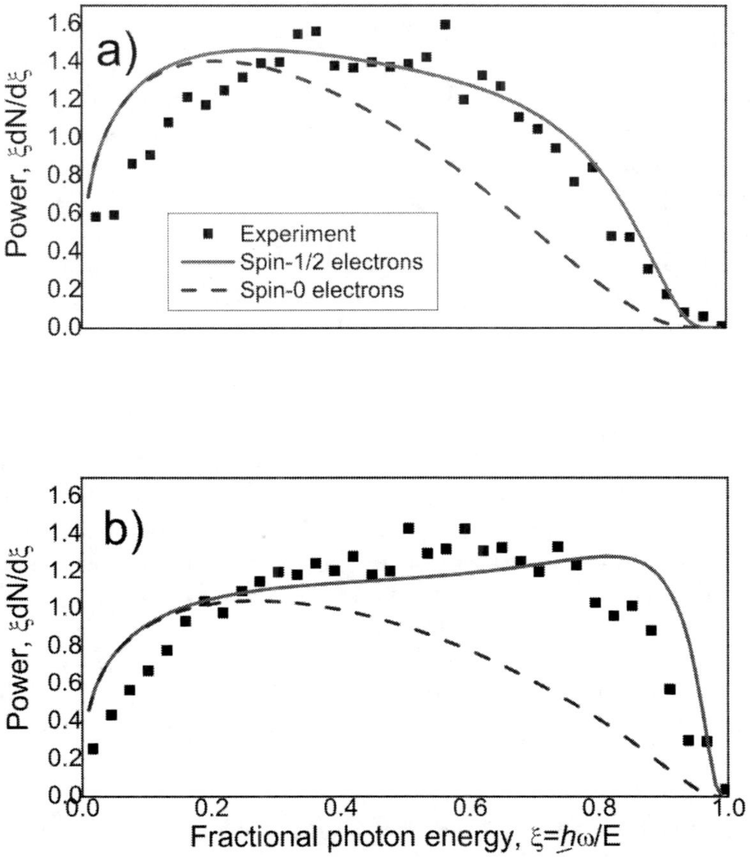

Figure 6. *Radiation emission as a function of photon energy for a) 70 GeV electrons and b) 243 GeV electrons. The full-drawn line shows the calculated radiation spectrum for spin one-half electrons at $\chi = 2.1$ and $\chi = 7.3$ (i.e. not averaged over the fields encountered, see also [36]), respectively, while the dashed line shows the corresponding quantity for scalar charged particles. The full-drawn lines can be compared to the dotted and dash-dotted lines in figure 1.*

terawatt laser pulses where non-linear Compton scattering and so-called Breit-Wheeler pair production are observed [37]. In none of these cases can the field be considered macroscopic.

From the foregoing discussion, it is clear that the two groups of Blankenbecler and Drell and Baier and Katkov have contributed substantially to the modern understanding of beamstrahlung. For the latter group, they - together with V. Strakhovenko - have been, and are still, world leaders in the theory of strong field effects in crystals. Both groups have furthermore contributed to other, related areas of recent experimental interest, as we discuss in short in the following.

6.1. Landau-Pomeranchuk-Migdal effect

The Landau-Pomeranchuk-Migdal (LPM) effect arises because of multiple Coulomb scattering that leads to a reduced radiation emission probability. The criterion for the onset of the effect is that the multiple scattering angle within the formation length must be larger than the Lorentz transformed $1/\gamma$ opening cone of radiation emission, such that emission amplitudes add destructively. The formation length is loosely speaking the distance of travel for the electron to separate by one reduced wavelength from the emitted photon ('the photon is set free'). For details on the formation length - also an extremely useful concept in strong field effects - the reader may consult [38, 39].

Blankenbecler and Drell developed their eikonal approximation methods to include the effect of a finite target for the LPM effect. This effect is known as the Ternovskii-Shul'ga-Fomin effect in radiation emission: that if the target becomes sufficiently thin, the radiation emission has the same dependence on photon energy as the normal amorphous (Bethe-Heitler) spectrum, only it is suppressed by up to an order of magnitude. Likewise, Baier and Katkov [40] increased the accuracy of calculations compared to the original approach by Migdal. In both cases, theory is in good agreement with measurements [41, 42]. It is clear that similar phenomena may appear in beamstrahlung, for sufficiently small bunch lengths, although this regime has not been touched upon in collider concepts so far.

6.2. Structured target resonances

As a continuation of the finite target approach developed with Drell, Blankenbecler enlarged the region of applicability of the theory by also considering resonances appearing from a stack of thin foils with constant separation, a so-called structured target (or sandwich target). It turned out that for such a target, resonance-like structures in the radiation emission spectrum appear for photon energies where the formation length is approximately equal to the foil separation. This calculation was confirmed shortly after by Baier and Katkov. Given a micro-structure in the colliding beams, such effects may also be present for beamstrahlung. Seen from an experimental viewpoint, however, the structured target resonances have proven very difficult to establish, since they amount to a $\simeq 10\%$ correction to the effect of one finite target.

7. Conclusion

We have shown the close connection between strong fields in crystals and in beamstrahlung. Crystals provide experimental data supporting calculations of QED processes in strong electromagnetic fields. This confirmation is important for the development of future electron/positron collider experiments, since simulations of the interaction region must rely on theory in order to calculate e.g. QED backgounds. Experimental studies of quantum synchrotron radiation, pair production and the trident process have been performed. All of them address important issues of strong field QED and all of them have confirmed the validity of the theory in the tested strong field regime relevant to for example the interaction region of CLIC.

8. Acknowledgments

We wish to thank the members of the NA43 and NA63 experiments at CERN who have all made an effort in obtaining the presented experimental data on crystals and amorphous targets. UIU thanks D. Schulte for enlightening discussions, K. Elsener for comments to the draft, and the organizers of the `Advanced QED methods for future linear colliders` workshop for an interesting programme and for the opportunity to present the experimental results.

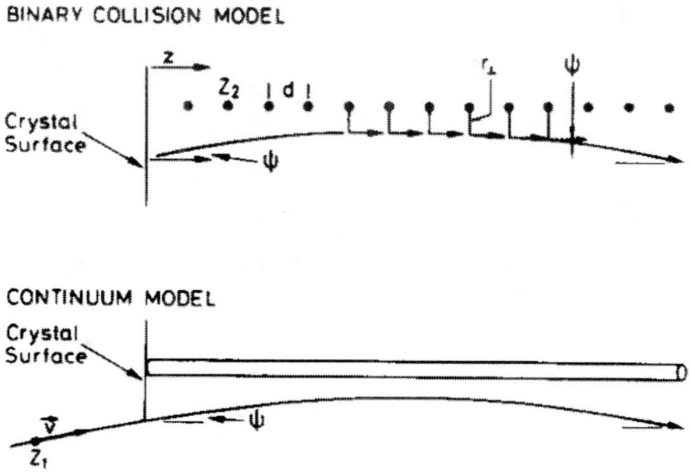

Figure A1. *A schematical drawing of the discrete nature of the scattering centers in a crystal and the resulting continuum approximation.*

Appendix A. Strong fields in crystals

In the following, a brief introduction to strong field effects in crystals is given with particular emphasis on pair production. However, essentially all conclusions and expressions can be applied also to the case of radiation emission and thus apply for the development of electromagnetic showers as well. For a more complete introduction to strong field effects in crystals, see [23], [34] and [43].

The large fields present near the nuclei in solids may in the case of single crystals add coherently such that a penetrating charged particle experiences a continuous field along its direction of motion, see figure A1. As a special case, if the particle is incident with a sufficiently small angle to a particular crystallographic direction, inside the so-called Lindhard angle, the negatively/positively charged particle is constrained to move near to/far from the nuclei and the electron clouds surrounding these. This is the channeling phenomenon [44] which has found widespread applications in physics. Another well-known example of coherent action is the emission of so-called coherent bremsstrahlung. This appears when crystal planes are crossed by the particle at regular intervals *and* at small angles to crystallographic directions during penetration [45]. The small angle requirement is essential for the coherence of the scattering that results in amplification of e.g. the radiation spectra.

On the other hand, at high energies (generally above a few GeV for electrons) a seemingly different coherent effect starts to appear. This effect arises due to the continuous field seen by the penetrating particle which acts on an electron with sufficient strength to impose a strongly curved trajectory (as compared to the energy - and thus typical emission angles - of the particle), see figure A1. The curvature - independent of the origin being magnetic or electric fields - gives rise to radiation emission of synchrotron character. Furthermore, this effect persists out to the so-called Baier angle, Θ_0, which is of the order of a few mrad and independent of energy

$$\Theta_0 \simeq Ze^2/dmc^2 \tag{A.1}$$

where d is the lattice distance along the axis. For particle energies above a few GeV (depending on crystal type), the Baier angle exceeds the Lindhard angle, and the particle is subjected to a continuously strong field even though it is not channeled, as long as it is incident within Θ_0.

The electric fields, $\mathcal{E}$, can be locally up to a few 10^{11} V/cm, depending on the crystal type and orientation. The incident particle moves in these immensely strong fields over distances up to that of the crystal thickness, i.e. up to several cm. For a sufficiently energetic particle with γ approaching 10^6, the critical parameter $\chi = \gamma\mathcal{E}/\mathcal{E}_0$ can thus reach values near and even beyond unity since $\mathcal{E}_0 = 1.32 \cdot 10^{16}$ V/cm. Thereby the behaviour of charged particles in macroscopic strong fields as $\mathcal{E}_0$ or B_0 can be investigated in accelerator based experiments.

It is essentially the same phenomenon which is responsible for the mrad deflection of multi-GeV protons [46] and ions [47] during the passage of a few cm of a bent crystal. In this case, the equivalent magnetic field is as high as several times 1000 T [48].

[1] Mohr P J, Taylor B N and Newell D B 2008 *Reviews of Modern Physics* **80** 633
[2] Fujiwara M *et al.* 2008 *AIP Conf. Proceed.* **1037** 208
[3] Jackson J 1975 *Classical Electrodynamics*
[4] VB Berestetskii E L and Pitaevskii L 1971 *Relativistic Quantum Theory* (Pergamon Press, New York)
[5] Schwinger J 1951 *Phys. Rev.* **82** 664–679
[6] Sauter F 1931 *Z. Phys.* **69** 742–764
[7] Sauter F 1931 *Z. Phys.* **73** 547–552
[8] Lindhard J 1991 *Phys. Rev. A* **43** 6032–6037.
[9] Baier V N and Katkov V M 2005 *Nucl. Instr. Meth. B* **234** 106–115
[10] Himel T and Siegrist J 1985 *SLAC-PUB* 3572
[11] Yokoya K 1986 *Nucl. Instr. Meth. A* **251** 1–16
[12] Noble R 1987 *Nucl. Instr. Meth. A* **256** 427–433
[13] Chen P and Telnov V I 1989 *Phys. Rev. Lett.* **63** 1796–1799
[14] Blankenbecler R, Drell S D and Kroll N 1989 *Phys. Rev. D* **40** 2462–2476
[15] Baier V N, Katkov V M and Strakhovenko V M 1988 *Zh. Eksp. Theor. Fiz.* **94** 125–139
[16] Blankenbecler R and Drell S 1987 *Phys. Rev. D* **36** 277–288.
[17] Solov'yov A V, Schäfer A and Hofmann C 1993 *Phys. Rev. E* **47** 2860–2867
[18] Solov'yov A V and Schäfer A 1993 *Phys. Rev. E* **48** 1404–1409
[19] Bell M and Bell J 1988 *Part. Accel.* **22** 301–306
[20] AA Sokolov N K and Ternov I 1952 *J. Exper. Theoret. Phys.* **23** 632
[21] Schwinger J 1954 *Proc. Natl. Acad. Sci. USA* **40** 132–136.
[22] Sokolov A and Ternov I 1986 *Radiation from relativistic electrons* (AIP, New York)
[23] VN Baier V K and Strakhovenko V 1998 *Electromagnetic Processes at High Energies in Oriented Single Crystals*
[24] Chen P and Yokoya K 1988 *Phys. Rev. Lett.* **61** 1101–1104
[25] Kirsebom K, Kononets Y, Mikkelsen U, Mller S, Uggerhj E, Worm T, Elsener K, Biino C, Doble N, Ballestrero S, Sona P, Avakian R, Ispirian K, Taroian S, Connell S, Sellschop J, Vilakazi Z, Moore R, Parker M, Baurichter A and Strakhovenko V 2001 *Nucl. Instr. Meth. B* **174** 274–296
[26] VN Baier V K 2006 *Phys. Lett. A* **353** 91–97
[27] Erber T 1966 *Rev. Mod. Phys.* **38** 626–659.
[28] Baier V and Katkov V 2005 *Phys.Lett. A* **346** 359–366
[29] Kirsebom K *et al.* 1998 *Nucl. Instr. Meth. B* **135** 143–148
[30] VN Baier V K and Strakhovenko V 1991 *Sov. J. Nucl. Phys.* **53** 632–637
[31] Kimball J and Cue N 1985 *Phys. Rep.* **125** 69–101
[32] Esberg J, Kirsebom K, Knudsen H, Thomsen H, Uggerhøj E, Uggerhøj U, Sona P, Mangiarotti A, Ketel T, Dizdar A, Dalton M, Ballestrero S and Connell S 2009 *in prep.*
[33] Baier V N and Katkov V M 2009 *Phys. Lett. A* **373**(21) 1874–1879
[34] Sørensen A 1991 *Vacuum structure in intense fields* **255** 91–118
[35] Kirsebom K, Mikkelsen U, Uggerhøj E, Elsener K, Ballestrero S, Sona P and Vilakazi Z 2001 *Phys. Rev. Lett.* **87** 054801
[36] AV Korol A S and Greiner W 2002 *J. Phys. G* **28** 627–641
[37] Bamber C *et al.* 1999 *Phys. Rev. D* **60** 092004.
[38] Baier V and Katkov V 2005 *Phys.Rep.* **409** 261–359
[39] Uggerhøj E and Uggerhøj U 2005 *Nucl. Instr. Meth. B* **234**
[40] Baier V N and Katkov V M 1998 *Phys. Rev. D* **57** 3146–3162
[41] Hansen H D, Uggerhøj U I, Biino C, Ballestrero S, Mangiarotti A, Sona P, Ketel T J and Vilakazi Z Z 2004 *Phys. Rev. D* **69** 032001
[42] Thomsen H D, Kirsebom K, Knudsen H, Uggerhøj E, Uggerhøj U I, Sona P, Mangiarotti A, Ketel T J, Dizdar A, Dalton M, Ballestrero S and Connell S 2009 *Phys. Lett. B* **672** 323–327
[43] Sørensen A and Uggerhøj E 1989 *Nucl. Sci. Appl.* **3** 147.
[44] Lindhard J 1965 *K. Dan. Vidensk. Selsk. Mat. Fys. Medd.* **34** no. 14, 1–64
[45] Palazzi G 1968 *Rev. Mod. Phys.* **40** 611–631.
[46] Baurichter A *et al.* 2000 *Nucl. Instr. Meth. B* **164-165** 27–43.
[47] Arduini G, Biino C, Clément M, Cornelis K, Doble N, Elsener K, Ferioli G, Fidecaro G, Gatignon L, Grafström P, Gyr M, Herr W, Klem J, Mikkelsen U, Weisse E, Møller S P, Uggerhøj E, Taratin A, Freund A, Keppler P and Major J 1997 *Phys. Rev. Lett.* **79** 4182–4185
[48] Møller S 1995 *Nucl. Instr. Meth. A* **361** 403–420

Journal of Physics: Conference Series **198** (2009) 012008 doi:10.1088/1742-6596/198/1/012008

Strong fields and recycled accelerator parts as a laboratory for fundamental physics

—

The quest for minicharged particles

Joerg Jaeckel

Institute for Particle Physics Phenomenology, Durham University, Durham DH1 3LE, UK

Over the last few years it has become increasingly clear that low energy, but high precision experiments provide a powerful and complementary window to physics beyond the Standard Model. In this note we illuminate this by using minicharged particles as an example. We argue that minicharged particles arise naturally in extensions of the Standard Model. Compatibility with charge quantization arguments suggests that minicharged particles typically arise together with a massless hidden sector U(1) gauge field. We present several low energy experiments employing strong lasers, electric and magnetic fields that can be used to search for (light) minicharged particles and their accompanying U(1) gauge boson.

1 Introduction

The development of high energy accelerators also leads to the development of technology to generate high electric and magnetic fields. In this note we want to explore how one can directly use these strong fields as laboratory for fundamental physics. High precision experiments in these strong fields allow us to search for new light particles which interact very weakly with the Standard Model particles. Due to their extremely weak interaction these hidden sector particles may be missed in conventional collider experiments. Accordingly the discussed high precision experiments provide complementary information on physics beyond the Standard Model.

In this note we will mainly discuss a particular type of hidden sector particle: minicharged particles. However, as we will see in many models minicharged particles are accompanied by an extra U(1) gauge boson, a hidden photon.

Minicharged particles interact with the ordinary electromagnetic field via the usual minimal coupling induced by the covariant derivative,

$$D_\mu = \partial_\mu - \mathrm{i}Q_f e A_\mu \tag{1}$$

where Q_f is the electric charge of the particle f. For example if f is a fermion the interaction term reads $Q_f e \bar{f} \not{A} f$.

The crucial point for a minicharged particle is now simply that the charge is much smaller than 1,

$$\epsilon = Q_f \ll 1. \tag{2}$$

In particular it is not necessarily integer. Indeed it does not even have to be a rational number. As we will see in the following section minicharges can arise in theories with kinetic mixing [1]

Journal of Physics: Conference Series **198** (2009) 012008 doi:10.1088/1742-6596/198/1/012008

or magnetic mixing [2]. Another possibility are scenarios with extra dimensions [3]. Typical predicted values, e.g., in realistic string compactifications range from 10^{-16} to 10^{-2} [3,4].

2 Minicharged particles in extensions of the Standard Model

Before we investigate how minicharges can be experimentally searched for in strong field experiments let us briefly show how they arise in simple extensions of the Standard Model. As we will see, in this simple setup the minicharged particles are accompanied by an extra U(1) gauge boson which also opens up new possibilities for experimental searches.

One way to have minicharged particles is in extensions of the Standard Model that contain an extra hidden U(1) gauge degree of freedom under which the Standard Model particles are uncharged. Such a model has a gauge group[1] $\mathrm{U}(1)_{\mathrm{QED}}\times \mathrm{U}(1)_h$. The most general renormalizable Lagrangian in the gauge sector is,

$$\mathcal{L} = -\frac{1}{4}\Big\{ \frac{1}{e^2} F^{\mu\nu}F_{\mu\nu} + \frac{1}{e_h^2} X^{\mu\nu}X_{\mu\nu} + \frac{2\chi}{ee_h} F^{\mu\nu}X_{\mu\nu} + \frac{1}{8\pi^2}\Big(\theta_{\mathrm{QED}}\, F^{\mu\nu}\tilde{F}_{\mu\nu} + \theta_h\, X^{\mu\nu}\tilde{X}_{\mu\nu} + 2\theta_\chi\, F^{\mu\nu}\tilde{X}_{\mu\nu}\Big)\Big\}. \tag{3}$$

Here, the $F^{\mu\nu}$ is the field strength of the electromagnetic $\mathrm{U}(1)_{\mathrm{QED}}$ with the gauge field A^μ and $X^{\mu\nu}$ is the field strength of the hidden $\mathrm{U}(1)_h$ with the gauge field X^μ. The gauge couplings are e and e_h, respectively. The dual field strengths are defined as usual:

$$\tilde{F}^{\mu\nu} = \frac{1}{2}\epsilon^{\mu\nu\kappa\lambda}F_{\kappa\lambda}\,, \tag{4}$$

and analogously for $\tilde{X}^{\mu\nu}$.

The standard kinetic terms as well as the θ_{QED}- and θ_h-terms are the diagonal entries in the Lagrangian (3). The CP violating θ terms are often neglected in standard electrodynamics because the corresponding terms correspond to total derivatives. However, in presence of magnetic monopoles the corresponding integrals do not necessarily vanish and the θ-terms become physically relevant (see below how this works for the non-diagonal θ_χ-term). Kinetic [1] and magnetic mixing [2] between the different U(1) sectors is represented by the off-diagonal χ-term and θ_χ-term, respectively.

It should be noted that both the kinetic mixing and the magnetic mixing term naturally arise via quantum corrections when heavy particles are integrated out [1,2] even when both U(1)s arise from a single non-abelian gauge group [2]. Indeed both χ and θ_χ are dimensionless and can therefore receive sizable contributions even if the particles involved are very heavy.

Let us now see how minicharges can arise in this setup by looking at two simple examples.
1. Kinetic mixing: Consider a situation in which all the θ-terms vanish but the kinetic mixing term $\chi \neq 0$. The kinetic terms in Eq. (3) can be diagonalized by a shift of the hidden gauge field,

$$X^\mu \to X^\mu - \frac{\chi e_h}{e}A^\mu. \tag{5}$$

Apart from a multiplicative renormalization of the gauge coupling,

$$e^2 \to e^2/(1-\chi^2), \tag{6}$$

[1]In principle one has to consider the mixing with the hypercharge $\mathrm{U}(1)_Y$, however at energies below the weak scale physics is well approximated by considering only the mixing with the electromagnetic $\mathrm{U}(1)_{\mathrm{QED}}$.

the ordinary electromagnetic gauge field A^{μ} remains unaffected by this shift. Consider now, for example, a hidden fermion f charged under X^{μ}. Applying the shift (5) to the coupling term, we find:

$$\bar{f}\not{X} f \to \bar{f}\not{X} f - \frac{\chi e_h}{e}\bar{f}\not{A} f. \tag{7}$$

Since the kinetic term is now diagonal, it is clear that the particle f (which was originally charged only under $\mathrm{U}(1)_h$) interacts with the $\mathrm{U}(1)_{\mathrm{QED}}$ gauge field and appears to have a charge $\epsilon = -\chi e_h/e$.

2. Magnetic mixing: Let us now look at the effect of the θ-terms. For simplicity let us take $\chi = \theta_{\mathrm{QED}} = \theta_h = 0$, but $\theta_\chi \neq 0$.

Let us assume we have a magnetic monopole under the hidden gauge group. In a static situation with such a hidden monopole as a background the electric and magnetic fields of $\mathrm{U}(1)_{\mathrm{QED}}$ and $\mathrm{U}(1)_h$ read

$$\mathbf{E} = \nabla A^0, \qquad \mathbf{B} = \nabla \times \mathbf{A}, \qquad \mathbf{E}_h = \nabla X^0, \qquad \mathbf{B}_h = \nabla \times \mathbf{X} + \frac{e_h g_h}{4\pi}\frac{\mathbf{r}}{r^3}. \tag{8}$$

The magnetic mixing part of the Lagrangian then reads

$$\begin{aligned} L_{\theta_\chi} &= \frac{\theta_\chi}{8\pi^2}\int d^3x\,(\mathbf{E}\cdot\mathbf{B}_h + \mathbf{E}_h\cdot\mathbf{B}) = -\frac{\theta_\chi}{8\pi^2}\int d^3x\, A^0\nabla\cdot\left(\frac{g_h e_h}{4\pi}\frac{\mathbf{r}}{r^3}\right) \\ &= -\frac{\theta_\chi e_h g_h}{8\pi^2}\int d^3x\, A^0\delta^3(\mathbf{r}) = -\frac{\theta_\chi}{2\pi}\int d^3x\, A^0\delta^3(\mathbf{r}), \end{aligned} \tag{9}$$

where we have used the quantization condition $e_h g_h = 4\pi$ in the last step (see next section). The last part is, however, exactly what we would expect for the interaction of a particle with an (ordinary) *electric* charge[2] $\epsilon = -\theta_\chi/(2\pi)$.

3 Minicharges and charge quantization

In the previous section we have seen that minicharges can arise quite naturally in extensions of the Standard Model with an extra U(1) gauge factor. However, one might wonder how this fits with the idea of charge quantization (for more details see [6]).

In the following we will see that the extra U(1) gauge boson plays an important role in the compatibility of minicharges and charge quantization arguments. In the main text we will focus on a variant of the charge quantization arguments based on angular momentum quantization [7] in the presence of monopoles. In the Appendix A we look at the original Dirac argument [8] and an argument based on the absence of black hole remnants (and which doesn't require magnetic monopoles).

Let us consider a configuration of a magnetic monopole with coupling g and an electric charge q with coupling strength qe. The angular momentum contained in the electric and magnetic fields of such a configuration is

$$\mathbf{L} = \int d^3x\, \mathbf{x} \times (\mathbf{E} \times \mathbf{B}) = \frac{qeg}{4\pi}\hat{\mathbf{n}}. \tag{10}$$

[2]This is a generalization of the Witten effect [5], where a magnetic monopole acquires an electric charge in presence of CP violation.

Journal of Physics: Conference Series **198** (2009) 012008 doi:10.1088/1742-6596/198/1/012008

Here, $\hat{\mathbf{n}}$ is the unit vector pointing from the electric charge to the magnetic charge, and the right hand side can be obtained by inserting the electric fields $\mathbf{E} = qe\mathbf{r}/(4\pi r^3)$ and the magnetic field $\mathbf{B} = g\mathbf{r}/(4\pi r^3)$ for the electric and magnetic monopole, respectively. Since the total field momentum vanishes the angular momentum is independent of the choice of origin. The quantization of angular momentum in quantum mechanics now requires

$$|\mathbf{L}| = \frac{qeg}{4\pi} = \frac{n}{2}, \tag{11}$$

where n is an integer (and as usual $\hbar = 1$). Enforcing Eq. (11) for our minicharged particle as well as an electron now requires that the ratio between the charges is rational, i.e. n/m with $n, m \in \mathbb{Z}$. This forbids irrational minicharges such as say $\sqrt{2} \times 10^{-12}$. However, in (3) the parameters χ and θ_χ are arbitrary real numbers. How can this be reconciled? The answer lies in the extra U(1).

So far we have considered only the ordinary electromagnetic field. However, if the minicharge arises as in the previous section via kinetic mixing[3] there is an additional U(1) gauge field. This field can now, too, contribute to the angular momentum. After the shift (5) the kinetic terms are diagonal and the generalized expression for the angular momentum is straightforward:

$$\mathbf{L} = \int d^3x\, \mathbf{x} \times (\mathbf{E} \times \mathbf{B} + \mathbf{E}_h \times \mathbf{B}_h). \tag{12}$$

With two U(1) factors we can, of course, also have more general magnetic monopoles. In general a monopole can have charges (g, g_h) under the visible and hidden magnetic fields. Its magnetic field will then be,

$$\begin{pmatrix} \mathbf{B} \\ \mathbf{B}_h \end{pmatrix} = \frac{\mathbf{r}}{4\pi r^3} \begin{pmatrix} g \\ g_h \end{pmatrix}. \tag{13}$$

We can now study static configurations of this monopole with:

a) an ordinary electrically charged particle (charge $q = 1$) with a field

$$\begin{pmatrix} \mathbf{E} \\ \mathbf{E}_h \end{pmatrix} = \frac{\mathbf{r}}{4\pi r^3} \begin{pmatrix} e \\ 0 \end{pmatrix} \tag{14}$$

and

b) a hidden sector particle (hidden charge $q_h = 1$) that has acquired a charge ϵ under the ordinary electromagnetic field,

$$\begin{pmatrix} \mathbf{E} \\ \mathbf{E}_h \end{pmatrix} = \frac{\mathbf{r}}{4\pi r^3} \begin{pmatrix} \epsilon e \\ e_h \end{pmatrix}. \tag{15}$$

Inserting into Eq. (12) we find

$$|\mathbf{L}_{a)}| = \frac{eg}{4\pi}, \qquad |\mathbf{L}_{b)}| = \frac{\epsilon e g + e_h g_h}{4\pi} = \frac{-\chi e_h g + e_h g_h}{4\pi}. \tag{16}$$

[3] Here, we will focus on kinetic mixing. For minicharges arising from magnetic mixing one can follow a similar strategy.

Angular momentum quantization now requires that both configurations have half-integer angular momentum,

$$|\mathbf{L}_{a)}| = \frac{eg}{4\pi} = \frac{n}{2} \quad \text{and} \quad |\mathbf{L}_{b)}| = \frac{\epsilon eg + e_h g_h}{4\pi} = \frac{-\chi e_h g + e_h g_h}{4\pi} = \frac{m}{2}, \tag{17}$$

where m and n are integers. It is clear that a naive monopole with $g_h = 0$ causes a problem because this would require $|\mathbf{L}_{a)}|/|\mathbf{L}_{b)}| = \epsilon = n/m$ and therefore ϵ to be rational. In the previous section we have, however, seen that ϵ arises directly from a parameter of the low energy Lagrangian and therefore it is typically not rational. This is the apparent contradiction produced by introducing both monopoles and minicharged particles.

However, a closer inspection of Eq. (17) reveals two types of monopoles which will not cause any such problems for arbitrary χ,

$$\begin{pmatrix} g \\ g_h \end{pmatrix} = \frac{2\pi m}{e_h}\begin{pmatrix} 0 \\ 1 \end{pmatrix}, \quad \text{and} \quad \begin{pmatrix} g \\ g_h \end{pmatrix} = \frac{2\pi n}{e}\begin{pmatrix} 1 \\ \chi \end{pmatrix}. \tag{18}$$

For the first monopole $|\mathbf{L}_{a)}| = 0$ and for the second $|\mathbf{L}_{b)}| = 0$.

In a pure U(1) setup we can simply choose the appropriate monopoles for consistency. If the U(1) arise from non-abelien gauge groups monopoles are unavoidable. When the symmetry is broken down to U(1)s they automatically arise as 't Hooft-Polyakov monopoles [9]. However, as it turns out [6] these 't Hooft-Polyakov monopoles have exactly the appropriate charges for consistency.

Above we have seen that minicharges arise naturally in one of the simplest possible extensions of the Standard Model, a Standard Model enlarged by an additional U(1) gauge group. More generally, this model can be viewed as a test model for a large class of Standard Model extensions which contain an unbroken extra U(1) gauge factor. Looking from a different angle, as we have seen consistency with charge quantization suggests that minicharged particles are generically accompanied by a (massless) U(1) gauge boson. This is not only of theoretical interest but will lead to interesting experimental tests. In the following three sections we will present tests for charge quantization that do not rely on the extra U(1) gauge boson before we make use of this feature in Sect. 7. Finally, in Sect. 8 we look at an experiment which is actually better suited to constrain minicharged particles *without* hidden photons, showing that one can experimentally distinguish between a situation with/without a hidden photon.

4 AC/DC an experiment to search for minicharged particles

Let us now turn to experimental searches for minicharged particles that employ (strong) electric and magnetic fields. The first experiment is directly sensitive to minicharged particles and does not rely on the existence of an extra hidden photon.

The basic setup for the experiment [10] is depicted in Fig. 1. In a strong electric field a vacuum pair of charged particles gains energy if the particles are separated by a distance along the lines of the electric field. If the electric field is strong enough (or the distance large enough) the energy gain can overcome the rest mass, i.e. the virtual particles turn into real particles. This is the famous Schwinger pair production mechanism [11]. After their production the electric field accelerates the particles and antiparticles according to their charge in opposite

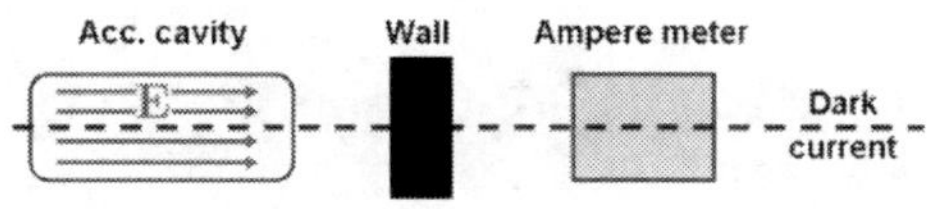

Figure 1: Schematic illustration of an *accelerator cavity dark current* (AC/DC) experiment for searching minicharged particles.

directions. This leads to an electric current (dashed line in Fig. 1). If the current is made up of minicharged particles the individual particles have very small charges and interact only very weakly with ordinary matter. Therefore, they can pass even through thick walls nearly unhindered. An electron current, however, would be stopped. After passage through the wall we can then place an ampere meter to detect the minicharged particle current.

Typical accelerator cavities achieve field strengths of $\gtrsim 25\,\mathrm{MeV/m}$ and their size is typically of the order of 10s of cm. Precision ampere meters can certainly measure currents as small as μA and even smaller currents of the order of pA seem feasible. Using the Schwinger pair production rate we can then estimate the expected sensitivity for such an experiment to be

$$\epsilon_{\mathrm{sensitivity}} \sim 10^{-8} - 10^{-6} \quad \text{for } m_\epsilon \lesssim \mathrm{meV}. \tag{19}$$

Therefore such an experiment has the potential for significant improvement over the currently best laboratory[4] bounds shown in Fig. 2.

Indeed, even without performing a dedicated experiment one can already obtain strong bounds from existing bounds on the energy loss in accelerator cavities [10]. The fact that high Q values have been achieved at large field values strongly limits the energy loss into unknown sources such as minicharged particle production. The corresponding bound is shown as the blue region in Fig. 2.

5 Tests of Coulomb's law as a probe for minicharged particles

Minicharged particles would also leave detectable footprints on the distance dependence of the electrostatic potential [15]. In presence of charged particles Coulomb's law is modified due to the effects of the vacuum polarization Π. This is the so-called Uehling potential [26] (for a textbook derivation see, e.g., [27]),

$$V(\mathbf{x}) = V_{\mathrm{Coulomb}}(x) + \delta V(x) = Q \int \frac{d^3q}{(2\pi)^3} \exp(i\mathbf{q}\cdot\mathbf{x}) \frac{e^2}{|\mathbf{q}|^2(1-\Pi_\epsilon(q))}. \tag{20}$$

For large distances, $r \gg 1/m$, δV drops off exponentially whereas for small distances, $r \ll 1/m$, it's behavior is logarithmic as expected from the running gauge coupling,

$$\begin{aligned} \delta V(r) &\approx \frac{Q\alpha}{r}\left[\frac{\alpha\epsilon^2}{4\sqrt{\pi}} \frac{\exp(-2mr)}{(mr)^{\frac{3}{2}}}\right] && \text{for} \quad mr \gg 1, \\ &\approx \frac{Q\alpha}{r}\left[-\frac{2\alpha\epsilon^2}{3\pi}\log(2mr) - a\right], \quad a \approx \frac{2\alpha\epsilon^2}{3\pi}\gamma && \text{for} \quad mr \ll 1. \end{aligned} \tag{21}$$

[4] Astrophysical bounds are much stronger [12] but are also somewhat model dependent [13].
[5] An interesting alternative to polarization experiments is interferometry [25].

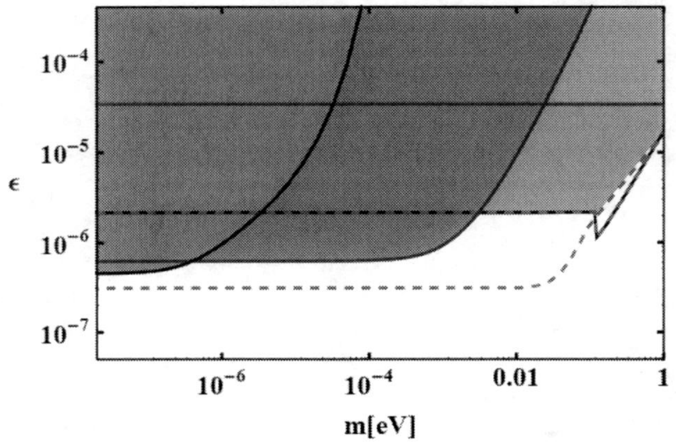

Figure 2: Laboratory bounds on minicharged particles. The black line (on the left, bottom solid) corresponds to the exclusion limit obtained in this note from the Cavendish type tests of Coulomb's law [14, 15]. The blue bound (on the left, middle solid) arises from constraints on energy losses in high quality accelerator cavities [10]. The dark green curve (on the left, top solid) gives the limit arising from bounds on the invisible decay on orthopositronium [16, 17] (a similar bound can be obtained from a reactor experiment [18]). The red-black dashed line denotes the limit [19, 20] arising from light-shining-through-a-wall experiments [21, 22] and applies only to minicharged particles arising from kinetic mixing whereas the red dashed curve gives a limit [20, 23] from polarization experiments [21, 24][4] and applies for a pure minicharged particle scenario. The shaded areas are excluded in both scenarios.

Due to their high mass electrons and heavier charged particles only lead to modifications of Coulomb's law at extremely short distances. In contrast light minicharged particles could lead to modifications of Coulomb's law at relatively large distances. We can therefore probe minicharged particles by testing Coulomb's law.

One way to test Coulomb's law is to use a so called Cavendish type experiment. Here, the electric potential between an outer charged conducting sphere and an uncharged inner sphere is measured[6]. A specific feature of Coulomb's law is that the electric potential inside a charged conductor is constant. Therefore, a non-vanishing voltage between the inner uncharged sphere and the outer sphere would indicate a deviation of Coulomb's law.

By applying high voltages to the outer sphere and looking for minuscle voltages between the inner and the outer sphere experiments of this type have been performed with enormous precisions of up to about 2 parts in 10^{16} for sphere sizes of 10s of centimeters [14]. This leads to the black bound shown in Fig. 2.

[6]Of course in reality the experimental setup is often more complicated and involves multiple spheres [14].

(a) (b)

Figure 3: **Left Panel:** Diagram depicting a classical process for a penetration of the barrier via conversion into a real particle that interacts only very weakly with the barrier. **Right Panel:** Diagram depicting "tunneling of the 3rd kind". The photon splits into a virtual pair of particle and antiparticle which traverse the wall and recombine into a photon.

6 Tunneling of the 3rd kind – Searching minicharged particles inside a superconducting box

Minicharged particles could also be searched for using a new type of tunneling process, "tunneling of the 3rd kind" [28]. This process is depicted in Fig. 3(b). Here, photons split into a *virtual* particle – antiparticle pair of charged particles which traverse an opaque wall and recombine after the wall into a photon. Due to their tiny interactions with ordinary matter the (virtual) minicharged particles can simply pass through the wall whereas photons would be stopped. After recombination the photons can then be detected. In this way minicharged particles lead to a light-shining-through-a-wall (LSW) signature. In contrast to the classical LSW process [29] (Fig. 3(a)) where the photon is converted into a single *real* particle the intermediate particles are *virtual*. Accordingly, this process depends on the thickness of the wall.

To observe this process we would like to use a relatively thin wall that is nevertheless completely intransparent to the photons themselves. For optical frequencies this is rather difficult (among other things due to the high energy of the photons). Therefore, here we will consider an experiment using static magnetic fields shielded by a superconductor. The basic idea [30] of the proposed experiment is very similar to a classic LSW experiments [29]. However, instead of light it uses a static magnetic field and the wall is replaced by superconducting shielding (cf. Fig. 4). Outside the shielding we have a strong magnetic field. Upon entering the superconductor the ordinary electromagnetic field is exponentially damped with a length scale given by the London penetration depth $\lambda_{\rm Lon}$. Yet, due to tunneling of the 3rd kind a small fraction of the magnetic field penetrates the wall and can be detected by a magnetometer. Since the magnetometer measures directly the field (and not some probability or power output) the signal is proportional to the transition amplitude and therefore to the charge squared, ϵ^2, instead of being proportional to ϵ^4.

High precision magnetometers can measure fields of the order of 10^{-13} T and even tiny fields of a few 10^{-18} T seem feasible. The expected sensitivity is of the order of $\epsilon \sim$ few $\times\, 10^{-7}$.

7 Searching minicharged particles with LSW experiments

In the previous section we have already encountered the possibility that minicharged particles can lead to a light-shining-through-a-wall signature. If there is also a light hidden U(1) gauge

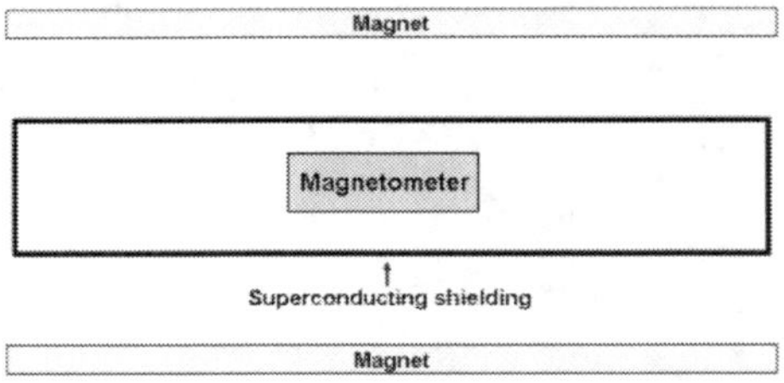

Figure 4: Sketched setup for the *superconducting box* experiment.

boson an additional process is possible and typically dominant: classical LSW via the hidden photon.

The essential process is depicted in Fig. 5(b). In presence of a non-vanishing magnetic background field the ordinary electromagnetic photon can be converted into a hidden photon via a minicharged particle loop [19]. The hidden photon can then traverse the wall and be reconverted on the opposite side of the wall via the inverse process. In terms of Fig. 3(a) the minicharged particle-loop is the (resolved) black dot and the hidden photon the dashed particle line.

Currently a sizable number of LSW experiments [21] are under way [22]. They employ strong lasers as light sources and recycled magnets from accelerator experiments to provide for the magnetic field. Their power lies in the fact that they can put in $\gtrsim 10^{20}$ photons per second and are often able to detect $\lesssim 1$ photon per second.

The bounds arising from LSW experiments are shown as the red-black dashed line in Fig. 2. Let us note again, that this bound on minicharged particles relies on the presence of an extra hidden U(1) gauge boson. But, as we have seen in the first few sections, this is quite natural.

8 Searching minicharged particles with laser polarization experiments

Finally, let us turn to laser polarization experiments [21, 24]. In these experiments polarized laser light is shone through a strong magnetic field and changes in the polarization are searched for. These measurements can be used as sensitive probes of minicharged particles [19, 20, 23].

We can have two different types of changes in the polarization: a rotation of the linear polarization and a change into an elliptical polarization. In particular the rotation is (nearly) free of Standard Model backgrounds and therefore very suitable to probe for new physics [31]. As shown in Fig. 6 a selective absorption of one polarization direction[7] via minicharged particle production results in a rotation of the linear polarization. A phase shift due to the vacuum polarziation leads to an ellipticity of the polarization.

If there are only minicharged particles Fig. 6 describes the whole story. For this case we obtain the red-dashed bound in Fig. 2.

[7] More precisely, photons of both polarization directions, parallel and perpendicular to the magnetic field, are absorbed, but the absorption coefficients differ.

Journal of Physics: Conference Series **198** (2009) 012008 doi:10.1088/1742-6596/198/1/012008

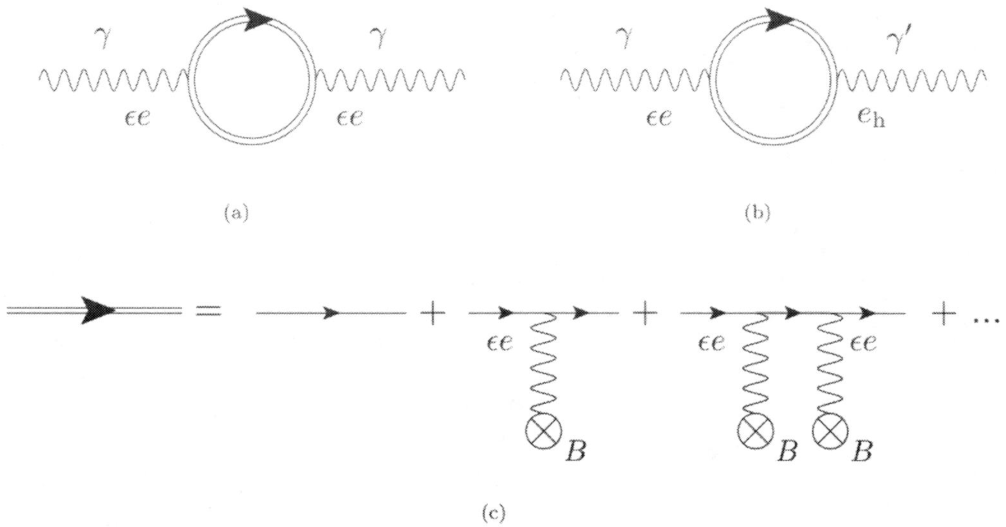

Figure 5: The contribution of minicharged particles to the polarization tensor 5(a). The real part leads to birefringence, whereas the imaginary part reflects the absorption of photons caused by the production of particle-antiparticle pairs. The analogous diagram 5(b) shows how minicharged particles mediate transitions between photons and hidden-sector photons γ'. Note that the latter diagram is enhanced with respect to the first one by a factor $\sim e_{\rm h}/(\epsilon e) = 1/\chi$. The double line represents the complete propagator of the minicharged particle in an external magnetic field B as displayed in 5(c).

In presence of an additional hidden photon the situation is more complicated. Indeed one has to solve the full equations of motion arising from the diagrams shown in Fig. 5 [19]. As it turns out in this case the bounds are typically weaker than those obtained with other methods described above. This can serve as an example that the extra hidden photon can also "hide" some of the experimental signatures of minicharged particles. More positively, this also explicitly shows that we can experimentally distinguish between the two situations.

9 Conclusions

Minicharged particles are a natural candidate for physics beyond the Standard Model. Their embedding into extensions of the Standard Model as well as charge quantization tests suggest that they are accompanied by an extra U(1) gauge boson, the hidden photon.

Light minicharged particles can be searched for in a large variety of laboratory experiments using high precision measurements in strong electric and magnetic fields. Using different experiments it is possible to distinguish between a situation which only has minicharged particles and one where the minicharged particle is accompanied by a hidden photon.

It should be mentioned that for minicharged particles astrophysical bounds [12] are often considerably stronger than the current laboratory bounds. However, astrophysical bounds are

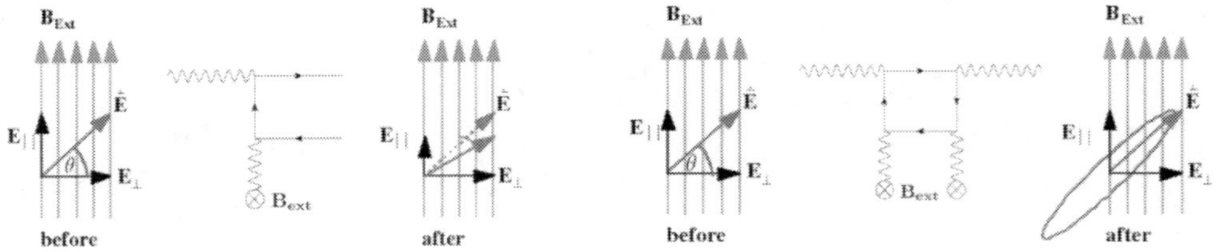

Figure 6: Rotation (left) and ellipticity (right) caused by the existence of a light particle with a small electric charge (figure adapted from [32]). More precisely, the Feynman diagram in the left diagram corresponds to the imaginary part of Fig. 5(a) and the one in the right diagram corresponds to the real part of Fig. 5(a).

also somewhat more model dependent [13]. Moreover, the recent rapid progress in the laboratory experiments raises the hope for considerable improvements in the near future.

Finally, the minicharged particles discussed in this note are only one example of a more general class of weakly interacting sub-eV particles (WISPs) that can be searched for in small scale laboratory experiments at low energies. Most of the above described experiments are also sensitive to other species of WISPs. For example, the extra hidden U(1) photon could also be massive. This leads to interesting effects such as photon – hidden photon oscillations even in absence of extra hidden matter that would become minicharged. For massive hidden photons experiments of the above described types provide the best bounds – even stronger than astrophysical bounds – for a large range of parameters.

With their small scale and rapid development low energy but high precision experiments inside strong elctromagnetic fields provide an interesting probe of physics beyond the Standard Model. They are not only complementary in the type of physics they probe but in addition they often recycle parts developed for accelerators such as high field magnets or high quality cavities.

Acknowledgements

The author wishes to thank the organizers of the workshop *Advanced QED Methods for Future Accelerators* for a stimulating meeting. Moreover he is indebted to S. Abel, M. Ahlers, F. Bruemmer, H. Gies, V. V. Khoze, J. Redondo and A. Ringwald for many many interesting discussions and fruitful collaboration on the subjects discussed in these proceedings.

A Alternative charge quantization arguments

In this appendix we will briefly consider alternative arguments for charge quantization and investigate how they are compatible with kinetic mixing.

A.1 A simple variant of the Dirac argument

In the main text we have used angular momentum quantization to argue in favor of charge quantization. Alternatively we can follow more closely Dirac's original strategy [8] and look at gauge transformations.

Let us take a look at the gauge field configuration of a Dirac monopole. The gauge field for a Dirac monopole with its string pointing in the $\mathbf{n}$ direction is,

$$\mathbf{A}^D(\mathbf{r}) = \frac{g}{4\pi}\frac{\hat{\mathbf{r}}\times\hat{\mathbf{n}}}{r(1-\hat{\mathbf{r}}\cdot\hat{\mathbf{n}})}, \tag{22}$$

where the hat denotes unit vectors.

A Dirac monopole with its string pointing in the north pole direction $\mathbf{n}_N = (0,0,1)$ is physically equivalent to one with the string pointing in the south pole direction $\mathbf{n}_S = (0,0,-1)$. Therefore, they must be related by a gauge transformation,

$$\mathbf{A}^D_N - \mathbf{A}^D_S = \nabla\Lambda. \tag{23}$$

Under this gauge transformation the wave function of a particle with charge q changes according to,

$$\psi \xrightarrow{\Lambda} \exp(\mathbf{i}qe\Lambda)\psi. \tag{24}$$

Since the wave function must be a proper continuous function $\exp(\mathbf{i}qe\Lambda)$ must be single valued.

Now let us calculate Λ for our gauge transformation given by Eq. (23). Moving the particle from $\mathbf{x}_0$ to $\mathbf{x}$ the gauge phase changes by,

$$\Lambda(x) - \Lambda(\mathbf{x}_0) = \int_{\mathbf{x}_0}^{\mathbf{x}} d\mathbf{s}\nabla\Lambda = \int_{\mathbf{x}_0}^{\mathbf{x}} d\mathbf{s}(\mathbf{A}^D_N - \mathbf{A}^D_S). \tag{25}$$

Now we can also imagine moving the particle around a closed loop. The single valuedness of the gauge transformation $\exp(\mathbf{i}qe\Lambda)$ then requires

$$2\pi n = qe\oint d\mathbf{s}(\mathbf{A}^D_N - \mathbf{A}^D_S), \qquad n\in\mathbb{Z} \tag{26}$$

for any closed curve.

For our gauge transformation changing the direction of the string from the north to the south pole the right hand side of Eq. (26) can be easily evaluated for a closed curve around the equator and we obtain the charge quantization condition,

$$2\pi n = qeg. \tag{27}$$

In the case of minicharges with kinetic mixing we can now ask ourselves how this quantization condition is fulfilled. As with the angular momentum argument the trick is to take into account the effect of the hidden magnetic monopole field. For the ordinary electron which has no hidden electric charge the argument works as above. However, for the minicharged particle arising from kinetic mixing Eq. (26) picks up an additional contribution from the hidden monopole field,

$$\begin{aligned} 2\pi n &= \epsilon e\oint d\mathbf{s}(\mathbf{A}^D_N - \mathbf{A}^D_S) + e_h\oint d\mathbf{s}(\mathbf{A}^{D,h}_N - \mathbf{A}^{D,h}_S), \qquad n\in\mathbb{Z} \\ &= \epsilon e g + e_h g_h = -\chi e_h g + e_h g_h. \end{aligned} \tag{28}$$

This is exactly equivalent to Eq. (17) and is automatically fulfilled if $g_h = \chi g$.

A.2 Charge quantization from the exclusion of black hole remnants

A completely different argument for charge quantization can be constructed by requiring that the theory is free of black hole remnants.

Let us assume for simplicity that we have only two types of particles with charge q_1 and q_2 with $q_2/q_1 \notin \mathbb{Q}$. Without restriction we can assume $q_1 > q_2 = 1$. For simplicity we assume that they have the same mass m. Because the charge ratio is irrational we can make arbitrarily small total charges by combining a number (n_1, n_2) or particles (and antiparticles)

$$1 \gg Q_{tot} = n_1 q_1 + n_2 q_2 > 0. \tag{29}$$

For each $\delta > 0$ we can now find a combination such that $|n_1| + |n_2| = N_{min}(\delta)$ is minimal and $\delta \geq Q_{tot} \neq 0$. Neglecting the (small) interaction energy the energy of such a configuration is $N_{min}(\delta)m$.

However, on the other hand $N_{min}(\delta)$ depends on δ and goes to infinity as $\delta \to 0$. Therefore we can construct a series $\delta_i > 0$ such that

$$\lim_{i\to\infty} \delta_i = 0, \qquad N_{min}(\delta_{i+1}) > N_{min}(\delta_i) \qquad \text{and} \qquad \lim_{i\to\infty} N_{min}(\delta_i) \to \infty. \tag{30}$$

The last statement means that with increasing i the energy of such configurations increases without bound.

Now, let us take such a configuration of charge $Q_{tot,i} \leq \delta_i$ and throw it into an uncharged black hole. Now this black hole can decay by emitting Hawking radiation of uncharged particles, e.g. photons, until it has reached an extremal Reissner-Nordstrom state[8] with

$$\delta_i^2 \geq Q_{tot,i}^2 = M_i^2/M_p^2, \tag{31}$$

and cannot decay any further without loosing charge. However, for sufficiently large i (cf. Eq. (30)), $N_{min}(\delta_i)m > M_i$ and the black hole also cannot shed charge by emitting charged particles because of energy conservation[9]. In other words the created black hole is stable. Indeed we have an infinite number of such remnant states in the theory. This is in violation of our assumption that the theory is remnant free.

Again we can ask ourselves if minicharges arising from kinetic mixing can circumvent this charge quantization argument. Indeed they can. The reason is again that the electrically minicharged particles also have a hidden electric charge. A combination of particles with a smaller and smaller visible electric charge then must have higher and higher hidden charges. This modifies the condition for the charged black hole,

$$M^2/M_P^2 \geq Q^2 + Q_h^2. \tag{32}$$

Due to the high hidden charge the black hole now typically has enough energy/mass to decay directly into charged particles and thereby shed its charge, allowing for a complete decay.

References

[1] B. Holdom, Phys. Lett. B **166** (1986) 196.

[8] Quantum gravity effects may already stop the decay at an earlier stage but this should not essentially change the following argument. We can always look only at sufficiently large i such that $N_{min}(\delta_i)m \gg M_P$.

[9] Emitting any number of particles smaller than $N_{min}(\delta_i)$ would only increase the total charge of the black hole and is therefore forbidden.

[2] F. Brummer, J. Jaeckel and V. V. Khoze, JHEP **0906** (2009) 037 [arXiv:0905.0633 [hep-ph]].

[3] B. Batell and T. Gherghetta, Phys. Rev. D **73** (2006) 045016 [hep-ph/0512356].

[4] K. R. Dienes, C. F. Kolda and J. March-Russell, Nucl. Phys. B **492** (1997) 104 [hep-ph/9610479]; A. Lukas and K. S. Stelle, JHEP **0001** (2000) 010 [hep-th/9911156]; D. Lust and S. Stieberger, hep-th/0302221; S. A. Abel and B. W. Schofield, Nucl. Phys. B **685**, 150 (2004) [hep-th/0311051]; S. Abel and J. Santiago, J. Phys. G **30**, R83 (2004) [hep-ph/0404237]; R. Blumenhagen, G. Honecker and T. Weigand, JHEP **0506** (2005) 020 [hep-th/0504232]; R. Blumenhagen, S. Moster and T. Weigand, Nucl. Phys. B **751** (2006) 186 [hep-th/0603015]; S. A. Abel, J. Jaeckel, V. V. Khoze and A. Ringwald, Phys. Lett. B **666** (2008) 66 [hep-ph/0608248]; S. A. Abel, M. D. Goodsell, J. Jaeckel, V. V. Khoze and A. Ringwald, JHEP **0807** (2008) 124 [arXiv:0803.1449 [hep-ph]]; G. Dvali and M. Redi, arXiv:0905.1709 [hep-ph].

[5] E. Witten, Phys. Lett. B **86** (1979) 283.

[6] F. Brummer and J. Jaeckel, arXiv:0902.3615 [hep-ph].

[7] H. A. Wilson, Phys. Rev. **75** (1949) 309; M. N. Saha, Indian J. Phys. **10** (1936) 145; Phys. Rev. **75** (1949) 1968; A. S. Goldhaber, Phys. Rev. **140** (1965) B1407.

[8] P. A. M. Dirac, Proc. Roy. Soc. Lond. A **133** (1931) 60; Phys. Rev. **74** (1948) 817.

[9] G. 't Hooft, Nucl. Phys. B **79** (1974) 276; A. M. Polyakov, JETP Lett. **20** (1974) 194 [Pisma Zh. Eksp. Teor. Fiz. **20** (1974) 430].

[10] H. Gies, J. Jaeckel and A. Ringwald, Europhys. Lett. **76** (2006) 794 [hep-ph/0608238].

[11] J. Schwinger, Phys. Rev. **82** (1951) 664.

[12] S. Davidson, S. Hannestad and G. Raffelt, JHEP **0005** (2000) 003 [hep-ph/0001179].

[13] E. Masso and J. Redondo, JCAP **0509**, (2005) 15, hep-ph/0504202; Phys. Rev. Lett. **97**, (2006) 151802, hep-ph/0606163; J. Jaeckel, E. Masso, J. Redondo, A. Ringwald, and F. Takahashi, hep-ph/0605313; Phys. Rev. **D75**, (2007) 013004, hep-ph/0610203; P. Brax, C. van de Bruck, and A.-C. Davis, Phys. Rev. Lett. **99**, (2007) 121103, hep-ph/0703243.

[14] E. R. Williams, J. E. Faller and H. A. Hill, Phys. Rev. Lett. **26** (1971) 721; D. F. Bartlett and S. Loegl, Phys. Rev. Lett. **61** (1988) 2285.

[15] J. Jaeckel, arXiv:0904.1547 [hep-ph].

[16] A. Badertscher *et al.*, Phys. Rev. D **75** (2007) 032004 [hep-ex/0609059].

[17] A. Ringwald, private communication.

[18] S. N. Gninenko, N. V. Krasnikov and A. Rubbia, Phys. Rev. D **75** (2007) 075014 [hep-ph/0612203].

[19] M. Ahlers, H. Gies, J. Jaeckel, J. Redondo and A. Ringwald, Phys. Rev. D **76** (2007) 115005 [arXiv:0706.2836 [hep-ph]].

[20] M. Ahlers, H. Gies, J. Jaeckel, J. Redondo and A. Ringwald, Phys. Rev. D **77** (2008) 095001 [0711.4991 [hep-ph]].

[21] R. Cameron *et al.* [BFRT Collaboration], Phys. Rev. D **47** (1993) 3707.

[22] K. Ehret *et al.* [ALPS Collaboration], hep-ex/0702023; C. Robilliard *et al.* [BMV Collaboration], Phys. Rev. Lett. **99** (2007) 190403; A. S. Chou *et al.* [GammeV (T-969) Collaboration], Phys. Rev. Lett. **100** (2008) 080402; P. Pugnat *et al.* [OSQAR Collaboration], arXiv:0712.3362 [hep-ex]. M. Fouche *et al.* [BMV Collaboration], Phys. Rev. D **78** (2008) 032013; A. Afanasev *et al.* [LIPSS Collaboration], Phys. Rev. Lett. **101** (2008) 120401; A. Afanasev *et al.* [LIPSS Collaboration], arXiv:0810.4189 [hep-ex]; K. Ehret *et al.* [ALPS Collaboration], arXiv:0905.4159 [physics.ins-det].

[23] H. Gies, J. Jaeckel and A. Ringwald, Phys. Rev. Lett. **97** (2006) 140402 [hep-ph/0607118]; M. Ahlers, H. Gies, J. Jaeckel and A. Ringwald, Phys. Rev. D **75** (2007) 035011 [hep-ph/0612098].

[24] E. Zavattini *et al.* [PVLAS Collaboration], Phys. Rev. Lett. **96** (2006) 110406 [arXiv:hep-ex/0507107]; S. J. Chen, H. H. Mei and W. T. Ni, Mod. Phys. Lett. A **22** (2007) 2815 [arXiv:hep-ex/0611050]; E. Zavattini *et al.* [PVLAS Collaboration], Phys. Rev. D **77** (2008) 032006 [arXiv:0706.3419 [hep-ex]].

[25] D. Boer and J. W. Van Holten, arXiv:hep-ph/0204207. G. Zavattini and E. Calloni, Eur. Phys. J. C **62** (2009) 459 [arXiv:0812.0345 [physics.ins-det]]; B. Doebrich and H. Gies, arXiv:0904.0216 [hep-ph].

[26] E. A. Uehling, Phys. Rev. **48** (1935) 55.

[27] M. E. Peskin and D. V. Schroeder, "An Introduction To Quantum Field Theory," *Reading, USA: Addison-Wesley (1995) 842 p*

[28] H. Gies and J. Jaeckel, arXiv:0904.0609 [hep-ph].

[29] L. B. Okun, Sov. Phys. JETP **56** (1982) 502 [Zh. Eksp. Teor. Fiz. **83** (1982) 892]. A. A. Anselm, Yad. Fiz. **42**, 1480 (1985); M. Gasperini, Phys. Rev. Lett. **59**, 396 (1987); K. Van Bibber, N. R. Dagdeviren, S. E. Koonin, A. Kerman and H. N. Nelson, Phys. Rev. Lett. **59**, 759 (1987).

[30] J. Jaeckel and J. Redondo, Europhys. Lett. **84** (2008) 31002 [arXiv:0806.1115 [hep-ph]].

[31] M. Ahlers, J. Jaeckel and A. Ringwald, Phys. Rev. D **79** (2009) 075017 [arXiv:0812.3150 [hep-ph]].

[32] F. Brandi *et al.*, Nucl. Instrum. Meth. A **461** (2001) 329 [hep-ex/0006015].

AUTHOR INDEX

Institute of Physics
Dirac House, Temple Back
Bristol BS1 6BE UK

CURRAN ASSOCIATES INC
proceedings.com